PRAISE FOR
NO-TILL INTENSIVE VEGETABLE CULTURE

"*No-Till Intensive Vegetable Culture* is the best practical manual for vegetable growers that I am aware of. From his deep wisdom and experience, Bryan O'Hara covers tillage alternatives, planting, seed saving, mineral balancing, mulching, composting, marketing, and all the other dynamics of running a successful vegetable farm. Far beyond numbers, techniques, and prescriptions, Bryan describes the expressions of principles and forces that govern growth, capturing and explaining subtleties and nuances that are foundational to vitality and providing readers with a deeper understanding that allows for better decision making."

—**Dan Kittredge**, founder and executive director,
Bionutrient Food Association

"Perhaps no calling demands so many disparate skills as that of the cultivation of vegetables to provide for others in harmony with nature. The compleat farmer is a master of all trades: observer, synthesizer, balancer of polarities, soil chemist, labor efficiency expert, coach, decision maker, harvester, marketer, and deep thinker. Using detailed examples and sharp tools, master farmer Bryan O'Hara has built an integrated philosophical framework and operations manual for this blessed work that has so much to offer both farmer and eater."

—**CR Lawn**, founder, Fedco Seeds

"Bryan O'Hara takes us on a journey to create a healthy, vibrant ecosystem that generates health through the production of nutrient-dense food while also helping to build community. I highly recommend this book to all those who seek a better future for humanity."

—**Gabe Brown**, rancher; author of *Dirt to Soil*

"Bryan O'Hara is one of the most brilliant vegetable farmers in North America. Few have mastered the ability to grow vegetables without diminishing the soil with destructive tillage. This humble soil genius understands the true art and science of vegetable farming. I highly recommend this book to all those who want to farm well and restore soil."

—**Ray Archuleta**, farmer; founder, Soil Health Academy;
partner, Understanding Ag LLC

"A must-read for anyone looking to take regenerative farming to the next level. When I visited Tobacco Road Farm, I realized that in a single day, I could only scratch the surface of all that Bryan O'Hara has to teach us. Whether it be Bryan's approach to no-till soil management, Korean Natural Farming and indigenous microorganism cultures, high-carbon composting, innovative mulching techniques, or year-round growing methods, every grower will find valuable ideas and instruction in this book. Bryan offers a whole-farm plan to grow productively while contributing to the healing of Mother Earth. Whether you follow Bryan's method in whole or in part, his ideas will go a long way towards helping you 'close the circle' and farm more ecologically, efficiently, enjoyably, and profitably while reducing outside inputs."

—**Andrew Mefferd**, editor, *Growing for Market* magazine; author of *The Organic No-Till Farming Revolution*

"*No-Till Intensive Vegetable Culture* is one of the most helpful guides for managing a vegetable production system I've encountered in over three decades of teaching in this area. Bryan O'Hara is a leading authority on building a farming operation that relies on knowledge rather than chemicals, and his concepts of balanced growth and crop plant polarities are innovative and insightful ways of looking at horticultural husbandry. Readers will gain intimate knowledge about working in an integrated manner with all components of a no-till production system. I take my students to Bryan's farm to experience his wellspring of innovative organic practices firsthand, and I plan on using this book as a text for my courses."

—**Dr. Gerald Berkowitz**, professor of plant science, University of Connecticut

"This book is a grower's dream, filled with crisp and invaluable details and clear-eyed explanations and guidance that connect 'how-to' with 'why.' O'Hara's innovative and pioneering methods amply demonstrate the power—and profitability—of prioritizing soil health and thriftiness alongside keen and insightful observations of field conditions. Simply put, you need this book at your fingertips if you want to successfully grow no-till vegetables."

—**Anne Biklé**, coauthor of *The Hidden Half of Nature*

"An inspiring guide to merging traditional and modern practices for successful small-scale vegetable farming based on rebuilding soil life. Having seen the incredible soil at Bryan O'Hara's farm, I can attest that it speaks for itself!"

—**David R. Montgomery**, author of *Growing a Revolution*

"Although I am not a no-till grower, I respect growers like Bryan O'Hara who are achieving great things using this method. *No-Till Intensive Vegetable Culture* is rich with intuitive detail; the text overflows with great understanding of soil and plants and how to cultivate them in a thoughtful, sympathetic, and beautiful way. To grow well, one must give of oneself, and Bryan O'Hara of Tobacco Road Farm has certainly done that. He is a true biodynamic grower, whose approach to the soil and its cultivation is assiduous. I have been a biodynamic grower for over twenty years, there is much for me to learn from O'Hara's book."

—**Jane Scotter**, Fern Verrow Farm, Herefordshire; coauthor of *Fern Verrow*

NO-TILL

Pesticide-Free Methods

INTENSIVE

for Restoring Soil

VEGETABLE

and Growing Nutrient-Rich,

CULTURE

High-Yielding Crops

BRYAN O'HARA

Chelsea Green Publishing
White River Junction, Vermont
London, UK

Project Manager: Sarah Kovach
Editor: Fern Marshall Bradley
Copy Editor: Laura Jorstad
Proofreader: Nancy Crompton
Indexer: Linda Hallinger
Designer: Melissa Jacobson
Page Composition: Abrah Griggs

Printed in the United States of America.
First printing January 2020.
10 9 8 7 6 5 4 3 2   21 22 23 24

Our Commitment to Green Publishing

Chelsea Green sees publishing as a tool for cultural change and ecological stewardship. We strive to align our book manufacturing practices with our editorial mission and to reduce the impact of our business enterprise in the environment. We print our books and catalogs on chlorine-free recycled paper, using vegetable-based inks whenever possible. This book may cost slightly more because it was printed on paper from responsibly managed forests, and we hope you'll agree that it's worth it. *No-Till Intensive Vegetable Culture* was printed on paper supplied by Versa Press that is certified by the Forest Stewardship Council.

Library of Congress Cataloging-in-Publication Data
Names: O'Hara, Bryan, author.
Title: No-till intensive vegetable culture : pesticide-free methods for restoring soil and growing nutrient-rich, high-yielding crops / Bryan O'Hara.
Description: White River Junction, Vermont : Chelsea Green Publishing, [2020] | Includes bibliographical references and index.
Identifiers: LCCN 2019044581 (print) | LCCN 2019044582 (ebook) | ISBN 9781603588539 (paperback) | ISBN 9781603588546 (ebook)
Subjects: LCSH: Vegetable gardening. | No-tillage. | Organic gardening.
Classification: LCC SB324.3 .O44 2020 (print) | LCC SB324.3 (ebook) | DDC 338.1/75--dc23
LC record available at https://lccn.loc.gov/2019044581
LC ebook record available at https://lccn.loc.gov/2019044582

Chelsea Green Publishing
85 North Main Street, Suite 120
White River Junction, VT 05001
(802) 295-6300
www.chelseagreen.com

FSC
www.fsc.org
MIX
Paper from
responsible sources
FSC® C005010

To increasing the abundance of life.

Contents

Introduction

The culture of vegetables has much to offer us. As growers we are continuing a co-evolution with the plants we call our vegetables. We assist them in bringing out the fullest expression of their potential, and they in turn help us bring out the fullest expression of ourselves. In this regard, to engage in the culture of vegetables at this time is a blessing. There is much to be grateful for when all that has led us down the path of growing these crops is considered.

This manual describes many of the practical details of growing vegetables in a beneficial relationship with nature. To give some context to how we have developed these methods, a description of our farm's evolution is perhaps a prerequisite. Anita and I started farming together at our present location in the early 1990s. We both had farming and gardening experience from our youth and were decided on making farming our means of livelihood from the beginning of our life together. With little money to start a farm, we first rented the small house and a few acres and began growing vegetables, employing some hand tools, a chain saw, a pickup truck, and a borrowed rototiller. The land had been severely depleted by past agricultural practices. It had started to reforest and needed extensive soil improvement. Fortunately, we had been introduced to the organic method of agriculture through our interactions with the health-oriented subculture of the 1970s and '80s. That, combined with our agricultural backgrounds, helped us understand the importance of compost and mineral application to improve soil conditions. Through the extensive additions of these materials and strong market conditions, we achieved early success, including clear demonstration that we could profitably produce vegetable crops without the use of pesticides. However, it was our friendship with our elderly neighbor Gilbert Risley that so quickly developed the farm into a successful endeavor. Gilbert had grown vegetables commercially for many years in the mid-1900s and had experience in everything from horse cultivation to chemical usage. Gilbert lent us a field behind his house, and every day he would come out and stand at the edge of the field, imparting to us his wisdom and knowledge. This invaluable mentorship continued until his death in 2001.

From Gilbert and other experienced growers, we quickly learned the details of traditional agriculture approaches to vegetable production. Our need to maintain a livelihood made us very attentive students of these farmers as well as of nature itself. With only worn-out soils and little money to rely on, this intensive study, along with much physical exertion and long hours, was required for many years to maintain the farm in a constant state of profitability. It is because of these conditions, which many non-enthusiasts would view as hardship, that Anita and I gained a thorough understanding of vegetable crop production.

With early success and profitability established, we quickly purchased the farm as well as invested in mechanization, including five old International tractors and a host of appropriate vegetable-growing implements. Since I had clearly caught "farming fever," my thought was to utilize the profits from a few acres of vegetable growing to amass the equipment and means needed to launch into a much larger farming operation. Luckily Anita was wise enough to break that fever, and I came to see that a few acres of vegetables was more than enough to maintain our livelihood and give us the life we sought. This life includes the health and happiness of our family, useful service to our fellow humans, spiritual devotion and growth, and significant independence and freedom from the modern system of economic subjugation.

The small farm operation that we had developed was providing well for these conditions, and with a clearheaded perspective came the realization that large-scale expansion would likely move us away from our primary intentions. Even 1 or 2 acres grown intensively in year-round production is sufficient to provide for our family's livelihood, though our present 3 acres (1.2 ha) of vegetables does give us more flexibility with the intensity of production and diversity of crops grown. For vegetable growing this size of operation is traditional worldwide, as this land area is an appropriate size for productive management by a family unit. Farmers and researchers from such diverse places as China and Ghana have told us that this is still the dominant size and mode of vegetable growing in their countries.

Over time, as environmental conditions in our local area deteriorated and we sought ways to improve the health of our crops, we were introduced to Biodynamics by aware gardeners in our community. Biodynamics, with its spiritual and holistic view of agriculture, was a natural fit for our developing farm. The initial applications of Preparation 500, also known as horn manure, resulted in a profound improvement in the aggregation of the farm's soils. The writings and transcribed lectures of Rudolf Steiner assisted us in integrating the spiritual world into our farming practices. The holistic perspective of Biodynamics gave us greater understanding of the interaction of natural forces with the web of life and thus improved our ability to grow healthful crops under challenging conditions.

Soon after we began Biodynamic practices, a farm intern introduced us to an agricultural methodology known as Korean Natural Farming (KNF), which she had studied while in Hawaii. This methodology improves soil function through the encouragement of biological diversity and through biological treatment of fertilizer materials. This approach put forth by Cho Han Kyu, who first developed KNF in the 1960s, utilizes techniques for cultivating the native biology from a region by collecting and multiplying microbes from nearby forests or fields to apply to farm soils. This involves various steps to culture the microorganisms and is very effective at producing large volumes of native biology to reintroduce into agricultural areas. The process is also known as indigenous microorganism (IMO) culture. Along with this process Cho describes many other approaches for the pre-fermentation or biological processing of various fertilizer materials in order to make them much more effective. Along with Biodynamics, practices of Korean Natural Farming have found much application in assisting crop production in the damaged environment that now faces us.

The study of agronomy has also proven useful in order to better understand soils and fertilizer usage. Soil testing, tissue testing, and a study of the chemistry of fertilizer interaction have helped us determine appropriate actions to increase the health of our crops. It has been particularly effective to combine information from scientific studies into our spiritual and holistic approaches, as these approaches can help guide each other to better agricultural outcomes.

Possibly the most important outcome from these studies and observations was the realization that tillage was having dramatic detrimental impacts on our soils. Soil aggregates were being pulverized, fungal organisms were obviously lacking in the field, and laboratory analyses of soil and plant tissue samples showed skewed nutrient profiles (such as too much nitrogen and potassium and not enough calcium, magnesium, phosphorus, and micronutrients). The final push that brought us to no-till, however, came from our practices of Korean Natural Farming. The emphasis on biological activity in that methodology highly discourages tillage, because it makes little sense to culture native microbiology through the IMO process and introduce it into the fields only to obliterate it by tilling. Over a period of about six years, we steadily reduced tillage and experimented with no-till methods until we were able to develop a system that was flexible and effective for intensive vegetable growing. Our switch away from tillage to no-till methods was nothing short of stunning in its improvement to soil and crop health, disease and insect resistance, weed control, irrigation reduction, labor reduction, improved efficiencies, improved crop storage, and more. Though for us no-till is just one aspect of our overall methodology, if one aspect had to be singled out as most dramatic in its influence it is probably safe to say that no-till was the one.

So after 30 years of growing at Tobacco Road Farm, we have seen much change in the world of vegetable production, and we have sought to change with it. All of our experiences and influences have allowed us to develop agricultural methods that have steadily maintained profitability through drought, hail, flooding, and other climate changes and manipulations. The soil remains strong, growing crops with excellent insect and disease resistance. There is no need to apply pesticides. The vegetables are rich in flavor and nutrients, and a family-friendly farming environment is maintained.

With our methods now solidly in place comes a desire to assist other farmers and gardeners in their efforts to grow nutritious foods for themselves. As growing conditions have steadily deteriorated, the establishment of a new, flexible, and nature-friendly agricultural methodology has become critical to success both economically as well as for maintenance of health. Our methodology has proven itself over many years of cropping on our farm as well as other farms in our region, often with impressive results in yields, quality, and profitability. This has led to many requests for a manual that could expand upon and provide a "solid copy" of the information presented at the lectures given on our methods, which invariably are too short to provide the full picture.

The production difficulties presently faced by our agricultural community are a result of the modern human condition and are unique challenges not faced by previous generations. These difficulties are numerous and constantly developing. The impact of pollution and human influence on air, sunlight, water, and soil quality is extensive and results in severe imbalances, which then lead to crop failures, often from pestilence. Fortunately for humanity, new farming methods are proving successful; these are the ones that work in a close harmonious relationship with nature. This relationship enlightens the agriculturalist's mind and functions as a teacher for not only improved agricultural practices but also lessons in the greater concerns of the human experience. The need for this experience is deeply felt in humans at this time as they seek a greater connection to the earth. As a result of this impulse, we are in a continuing back-to-the-land movement. The benefit of the methodology presented here is that it readily puts a livelihood into the hands of the people feeling this impulse, as it requires a relatively small land base of a few acres or less and little capital investment in mechanization. These concepts and techniques have been particularly attractive to the large number of both women and men who are presently entering into farming—which is not to say that these methods are not useful to both gardeners and established farmers as well. This methodology is based upon holistic concepts and observation of the whole of the environment, with its web of interconnection, and how this web interacts with vegetable growth. The holistic approach is certainly the traditional approach to agriculture. However, in order to present these holistic approaches in book format they, of necessity, must be somewhat taken apart and examined in sequence, only to be put back together again in an attempt to show the holistic picture. Hopefully I have succeeded in this endeavor.

The methods presented in this book are relatively complex and interconnected. They provide growers with no-till, pesticide-free, high-yielding, efficient practices, and they are described in detail so growers can understand why they are useful and how they influence the growing environment. These methods are successful because they are interconnected; actions rely on and assist other actions. Growers may do well by carefully following these methods. However, the primary objective of this manual is to help growers formulate a set of actions that may be best in their own environments, and in that regard some methods described herein may be more appropriate for adoption than others. As such, this manual is meant to develop growers' abilities for their own situations, but it is certainly not the last word on vegetable-growing technique. A manual is a picture in time, and the words on the page cannot evolve on their own. However, farms are continually in a state of development, and my hope is that many growers can take the information presented in this manual and develop it even further in their fields and gardens. These further developments are what will continue to provide us with the ability to feed ourselves at the same time as we gain a closer relationship with the natural and spiritual world. May the lessons we learn together provide us with inspiration, devotion, and gratitude to continue our work and allow us to emanate our understanding to our fellow humans, and the world.

CHAPTER 1

The Growing Environment

The four elemental states that are of primary importance for vegetable production are soil, water, air, and sunlight (warmth). Appropriate management of these elements in concert with one another leads to a successful crop. Therefore it is of great benefit to establish a farm or garden site where these states are naturally in relative balance. This can reduce management efforts and potential grower errors. There will always be year-to-year imbalances, however, or the need to produce crops on less-than-perfect lands, and thus it is important to develop a variety of techniques to reestablish balance.

The four elements are another way of expressing the various states of matter—solid, liquid, gas, and energy—or the more esoteric concepts of physical manifestation: earth, water, air, and fire. Let's start with a look at the most solid form: earth or soil.

Soil

Though the four elemental states all require attention and assistance to provide the best growing conditions for vegetable crops, soil is the traditional area that growers focus on most, because soil improvements are often long lasting and readily observable. Abundant healthful crops are the direct result of a grower-assisted, highly functioning soil with air, water, and sunlight provided through natural conditions. Thus, it is critical for growers to consider their actions in terms of whether they are of benefit or detriment to the evolution of soil function. Tillage and application of pesticides are areas of particular detriment. Though in the short term or under specific conditions these activities might be necessary, no-till, pesticide-free vegetable growing highly supports the efforts to improve soil.

Soils are often primarily formed from the decomposition of the parent rock material that underlies a region. Various forces such as weathering, chemical reactions, and biological activities break up this rock into smaller and smaller fragments. As the bedrock fragments it breaks into boulders, then into pieces of smaller size from fieldstones to cobblestones to pebbles, sand, silt, and clay. These mineral fragments are largely responsible for the structure of the soil and have a great influence upon soil's interaction with the elements of water, air, and warmth. A pebbly sand will allow much air and water to infiltrate, and it will warm in the sunlight quickly. This can be of benefit in a cool rainy environment but a bane under hot, dry conditions. Vice versa, a solid clay soil will sit cold and wet with little air penetration yet may perform very well during an extended sunny, dry period. These are simple examples, but understanding the effects of soil structure is of paramount importance in knowing how to handle crop production on specific soils. This is assisted by thoroughly understanding the physical characteristics of the soils and the conditions that led to their formation. This generally requires a deep dig into the soil to determine the makeup of its layers—the topsoil and subsoil—and perhaps also the depth of these layers to bedrock.

Soil survey maps from the USDA Natural Resources Conservation Service show the makeup of the soils in a given area and are generally quite accurate. They are available at libraries or on the internet.* Useful as these maps are, they are not a substitute for direct observation of the soil. For this, it is necessary to dig a hole at least 2 feet (60 cm) deep. It is possible to then examine the ability of air and water to penetrate through the layers and see the organic matter distribution in the soil profile. The texture of the various layers can be felt between the fingers to assess clay, silt, sand, and stone content. The soils can be smelled for conditions of anaerobic or aerobic activity. Depth of rooting and depth of earthworm activity are also indications of how well a soil is functioning. This kind of assessment provides critical basic knowledge for further decisions regarding the management of the elemental states. Some soil testing laboratories also provide information on soil texture and structure in their reports, and this may be of use as well. (More on soil testing in chapter 8.)

In addition to the mineral fragments, there is also organic matter incorporated into the soil profile. Organic matter is the carbon-containing material of life. For the purpose of this book, *organic* is defined as carbon-containing materials derived from living organisms, and use of the term does not necessarily denote materials or methods that conform to the USDA's National Organic Program standards ("certified organic"). The soil life gathers its living mass by recycling organic materials, as well as by taking up minerals released by the rock

---

* Go to https://websoilsurvey.sc.egov.usda.gov/App/WebSoilSurvey.aspx to find the soil survey maps.

fragments along with water and air in processes fueled by the sun. Once life has incorporated non-living mineral material into itself, much of this mineral material does not readily return to a non-living form. Instead life seeks to keep this material in the realm of life. The now biologically active materials are constantly recycled and built upon to create greater conditions of life. This yields the complexity and diversity that we perceive as the beauty of nature.

This biological recycling of organic forms is concentrated in two areas in the soil: around decaying organic materials and around the living roots of plants. Growers provide for these conditions of concentrated activity by supplying the soil with decomposable organic materials and covering the soil with growing plants as much as possible. In a symbiotic relationship, living plants actively supply carbon-rich nutrients to the soil organisms through their root exudates, and the soil organisms in turn supply the plants with nutrients derived from the soil environment. Much of this exchange is in the form of large, complex organic molecules, which are of great benefit to both. This saves the energy needed for synthesis of these complex molecules. The conserved energy can then be utilized by the plant in other areas of growth or development. This efficiency has extensive ramifications for the overall vitality of crops. The plants can supply the soil life with sugars and other carbonaceous materials from active photosynthesis. The soil life can supply the plants with mineral materials synthesized into organic forms by their metabolism. This symbiotic relationship is of paramount importance in the raising of well-balanced, vital crops, and growers are well rewarded for assisting its capacity.

The area around plant roots is referred to as the rhizosphere. In a functioning rhizosphere, the plants and soil life exude a gelatinous substance that creates conditions that cause soil particles to stick to the roots. Growers can observe these soil-coated roots by uprooting a plant that is growing in biologically active soil. The rhizosphere provides for not only a concentration of bacterial activities but also an appropriate environment for the connection to the fungal realm. The fungal organisms, including those termed mycorrhizae, form bonds and symbiotic relationships in this area that are of particular

FIGURE 1.1. Root exudates have created conditions that cause soil to adhere to the roots of these young peas. Note the abundant rhizobial nodules.

use to the plants. Fungal mycelia can reach out into the soil to a far greater extent than plant roots alone. These fungal organisms are able to supply plants with materials such as phosphorus, calcium, and many micronutrients that are otherwise difficult to assimilate into living forms.

During the process of organic matter recycling, some of the organic materials are biologically formed into long carbon chains called humic substances. These complex molecules are very stable forms of carbon in the soil and aid the ability of life to proliferate by supplying a foundation for nutrient absorption. This spongelike material is able to keep nutrients available for use in living systems.

Nature is constantly using the dynamics of the living to bring further life into soils by converting inorganic materials to organic materials. Growers are given the chance to aid nature in these activities and bring more life to our earth. This gives human life much meaning, and also provides plenty to do.

Air and Water

The combination of rock fragments and organic material makes up the solid particles we call soil. Of equal importance is the space between these soil particles—the pore spaces where air and water interact with each other and the soil. It is very important for growers to strive to create conditions of balance between the soil particles and the pore spaces. Generally this entails management choices that lead to the formation of aggregates or crumbs in the topsoil, which goes a long way toward allowing air and water, as well as warmth, to effectively penetrate the soil. The aggregation of soil is due to the effective decomposition of organic materials and the resulting biological "glues" secreted by the soil life that hold soil particles together. Growers can manage residue decay and apply mulches and composts to assist in this soil development. Worm castings are an easy-to-spot example of the soil life's attempts to aggregate soils. Lack of soil aggregation can lead to excessive soil compaction. Growers can address the problem of soil compaction, both surface and subsurface, through appropriate forms of tillage or non-tillage, organic mulches, compost, cover crops, fertilizers, and biological materials, as well as by taking care with machine traffic. Although management choices differ depending on soil type and condition, in general a highly productive soil is well aggregated on the surface and to some depth into the topsoil. It has no compacted layers or "plow pans" below, and thus the surface does not crust over and there is a relatively homogeneous structure through the deeper soil.

Well-aggregated soils are rarely found in commercial vegetable fields. Often there is both a crusted surface on the soil and compacted layers or plow pans below as a result of tillage and machine traffic. Excess tillage pulverizes aggregates, leading to a collapse of soil structure. Air cannot readily penetrate

FIGURE 1.2. The soil in this post-harvest conventional corn field near Tobacco Road Farm has a crusted surface and is actively eroding.

the crusted soil, thus reducing the ability of the aerobic (oxygen-dependent) microbes to proliferate. These are the microbes that lead to abundant, vigorous yields, but in their place anaerobic organisms may well proliferate, which in excess can result in a buildup of toxic gases in the pore space. These toxic gases cannot be released from the soil due to the crusted soil surface. The outcome is vegetables that suffer poor growth and are prone to disease and insect assault. When air can readily penetrate, the soil breathes in a steady rhythm, inhaling gaseous oxygen as well as nitrogen gas, which aerobic organisms need for respiration and nitrogen fixation. A porous soil also exhales the byproducts of respiration, including the carbon dioxide produced by aerobic organisms. This carbon dioxide is exhaled directly into the space just below the canopy of growing plants, where the leaves await this critical component of photosynthesis. This cycling is all appropriately timed by nature in a delicate perfection where the soil exhales at the time of day when the stomata are most likely to be open to allow the gas to be absorbed. A highly functional soil also cycles the proliferation of aerobic and anaerobic organisms at appropriate times for effective, diverse nutrient release. So much of nature is finely balanced, and thus the imperative to treat soils carefully and allow nature to function in a manner it has adjusted itself to. In other words, as a grower, be careful not to get in the way of delicate, naturally functioning systems.

Undisturbed homogeneous soil has a spongelike nature and generally has the ability to move soil water up from its depths. This is commonly seen in forest soils, which are able to remain moist for long periods of time during drought. In agricultural fields, plow pans or compacted layers caused by the pressure of equipment driving over the soil or the action of tillage tools inhibit such water movement into the topsoil; they prevent water in deeper layers from moving upward. A crusted soil surface also inhibits movement of water, in this case from rain or irrigation, down into the soil. This can lead to runoff and erosion as well as a lack of moisture for crop and soil life.

The naturally sustained balance of water and air in soil pore spaces is of the greatest importance to a steady reliable nutrient release from the soil life, so critical for vibrant, healthful crops. It is beneficial for growers to seek to mimic the natural conditions of undisturbed soil in their growing areas. This would take the form of a layer of undecomposed organic matter on the soil surface, more fully decomposed organic matter (such as a layer of compost) beneath, a topsoil undisturbed by tillage, and below that a subsoil without the presence of a compaction layer. Water can then readily be absorbed into the soil as well as move upward when soils begin to dry. If soils are allowed to dry out, a biological crash occurs and a diminishment in the soil life ensues. This soil life cannot instantly renew itself upon moisture recharge. It takes time for the microbes to

FIGURE 1.3. Notice the decomposing mulch at the surface of this no-till field soil and the aggregates formed underneath.

reestablish balance, all to the detriment of a crop. This type of biological crash is a leading cause of crop failure in the summer season. When disturbed soils dry excessively, effective biological nutrient delivery then falters and the resultant nutrient-deficient crop is assaulted by insects and disease. Anytime growers disturb the soil's air and water balance through tillage, excessive irrigation, inappropriate fertilization, or other practices, they may create such a dramatic upset of balance that nutrient imbalances manifest in crops.

The air *above* the soil surface is also of importance when considering crop production. The carbon dioxide in ambient air is the major source of carbon for crop growth. As noted above, plants inhale air through their leaf stomata, and the carbon from carbon dioxide is incorporated into the sugars that are the end result of photosynthesis. Management of air is also related to how the wind moves air across the growing area. Very windy conditions cause plants to "hunker down" and keep the canopy compact to resist damage. In very still conditions, however, plants exhibit more expansive, though potentially weaker, top growth. The wind condition of a farm is related to the lay of the land, and growers will benefit from a study of wind conditions across the growing area in order to understand air movement patterns. Usually there is a dominant direction of wind movement, often from the west. When conditions are windy, observe the variations in how plants are moving. This often occurs around windbreaks, whether they be trees or buildings or planned windbreak structures. The way in which air moves across land is very similar to the flow of water, in particular when it meets an obstruction. Thus a study of water movement will greatly aid in the understanding of air movement.

Windbreaks can be of great benefit because excessive air movement across the growing area can be detrimental. It may cause a loss of moisture or additional cooling that is unwanted, especially for crops grown during cold seasons. Windy conditions also make it difficult to keep row covers and hoop tunnels in place. As a general rule, every 1 foot (30 cm) of windbreak height results in 10 feet (3 m) of horizontal wind protection. Thus, a planting of 10-foot-tall trees offers 100 feet (30 m) of wind protection to the leeward side (deciduous trees offer less protection when they are not in leaf, though).

Trees and shrubs are common windbreak plants, conifers particularly being useful for cold-season protection of vegetables. The impact of a windbreak on sunlight levels needs to be considered, as well as any possible competition between the tree roots and vegetable crop roots. However, these roots may be symbiotic to some degree, especially due to tree roots' ability to maintain mycorrhizal fungi and other soil microbes during breaks in the vegetable cropping cycles.

Another choice for windbreaks is wide-webbed plastic snow fencing. Wooden slat fencing is another possibility. Like trees, these fences break up the wind without completely stopping it. They are more effective at slowing wind than is a solid

FIGURE 1.4. A winter day in the home field at Tobacco Road Farm. The windbreak cuts down on the dominant winter wind and helps keep the crops growing under the low tunnels warmer.

barrier, which may create a more turbulent flow of air as the wind is forced up and over the barrier. Solid barriers may well be better than nothing, however; stacked hay bales, solid wooden fencing, and buildings are examples of solid windbreaks.

Sunlight

The primary provider of energy for the growing system is of course the sun. The sun provides plants with the energy for the photosynthetic reaction, which, as mentioned above, combines carbon dioxide and water to create the simple sugar glucose and oxygen gas. Glucose provides the building blocks for further reactions that produce more complex molecules as well as providing an energy source for these reactions. As growers we assist in this by encouraging crop leaves to be well positioned to collect solar energy. This management of the crop canopy is an essential aspect for maximum health of a crop and the soil environment. The soil environment responds well to full coverage by a growing plant canopy. The canopy provides physical protection from the drying effects of wind and sun as well as providing the sugars and nutrients from plant root exudates to maintain active soil biology. In an overly dense canopy, however,

FIGURE 1.5. These vibrant, healthy rows of kale provide complete canopy coverage, protecting soil biology.

insufficient sunlight reaches each individual plant, which reduces photosynthesis and thus vitality and health of the individual. Conversely, when the canopy is too thin, soil can suffer from exposure and lack of plant root exudates, and the microclimate under the canopy may not be effective for carbon dioxide and water retention. Vegetable growers fulfill this primary duty to maintain an appropriate plant canopy through attention to reducing tillage impacts, timing seeding dates of vegetable crops appropriately, paying attention to cover crop seeding rate and crop spacing, and timely transplanting.

Tillage obviously destroys any plant canopies in place, with the goal generally being to create an open soil surface that can be easily planted with crop seeds or transplants. Tillage therefore has a tremendous impact on a soil's ability to feed and protect itself with a cover of growing plants, and reducing tillage can be of great benefit in order to lessen this deleterious impact. This may mean a no-till approach, strip tilling, or other techniques where minimal tillage is performed carefully and does not result in long fallow periods. In general a period in which soil is sitting barren is a period of loss of soil fertility and profitability.

Growing a diversity of vegetable crops is very helpful in the pursuit of keeping soil covered by a living canopy. Vegetable crops that form canopies fast are beneficial in mixed plantings. The role of annual plants in nature is often to cover soils quickly after a disturbance, and many vegetables can fill this role because they are fast growers. For example, red radish can be grown alongside slower-growing vegetables like celeriac. When the speedy crops are harvested and removed, the growing area is then opened up for the slower-growing crop. Another approach is to seed fast-growing annuals alongside sprawling crops. An example of this is seeding arugula or radish alongside cucumber, melon, squash, or tomato. The arugula or radish reaches maturity after about 40 days and the crop is harvested, and the sprawling crops are then allowed to grow into the harvested area. Cover crops can be seeded alongside a crop in a similar fashion and then either terminated, or not, depending on conditions.

Intercropping is another example. In this case, a second crop is seeded or planted alongside or underneath an initial crop that is already growing well or is reaching maturity. Examples of this include broadcast seeding leafy greens under sweet corn and broadcast seeding a cover crop into a vegetable crop. Interseeding can greatly perpetuate soil coverage with living plants and is very beneficial but does require that a thorough weed control program is in place. Broadcast seeding is usually the best technique when interseeding, because the extent of the existing plant canopy often limits the practicability of using a mechanical seeder (more about this in chapter 5).

The timing and methods of seeding and planting are worthy of consideration in terms of crop canopy coverage. Most crops grow more quickly during the summer, for example, which has an impact on choice of seeding date. Dry conditions in summer may result in poorer germination, however, so seed rates may have to be increased above the usual amount. Furrow seeding may be superior under dry conditions, yet broadcast seeding may result in better coverage if conditions are moist, as with seeding before rain or irrigation.

The physical spacing of a crop is of primary importance in the development of a beneficial canopy. In achieving an effective crop canopy, there is some flexibility. Many crops can tolerate a little crowding, and each individual plant will be smaller than typical. If planted at wide spacing, each individual will be larger. Transplanting gives a higher level of control of crop spacing than does seeding. As long as the plant canopy develops to fully cover the soil, a beneficial microclimate is created. Of course, transplants generally take less time to develop an effective canopy than does a seeded crop. Transplanting can also be very effective for canopy continuity because transplants can be interplanted into a crop soon to be harvested.

Cover crops, either interseeded or following a harvested crop, also contribute to canopy maintenance. Many cover crops such as annual grains or buckwheat are vigorous growers and can provide coverage quickly. Others, like some clovers, can tolerate the shade when they are undersown below a growing crop.

FIGURE 1.6. These new melon plants (*above*) are growing alongside a cover crop. After the cover crop is terminated, the melons then sprawl into the open space (*below*).

FIGURE 1.7. A field pea cover crop fills the space between two beds of young tomato plants. Later on the peas will be crushed down by foot and mulched to terminate them.

Even weeds can serve as an effective canopy, often quickly! Of course they usually will need to be terminated before they set seed.

In order to discern the appropriate crops for a given time and place, growers need to be aware of the levels of light and warmth available over the course of the growing season in the various areas of fields. The sun's energy in the form of light is interrelated with the force of warmth. Often the conditions that lead to high light exposure will increase the warming of a given area. Day length is an example: The warmest time of year usually occurs just after the longest day of the year (summer solstice), and the coldest time shortly after the shortest day of the year (winter solstice). The slope of the land also impacts light exposure and warming. In northern latitudes a south-sloping field will gain more warmth and aid photosynthesis more than other slopes, particularly in winter when the sun's angle is lower. Hilltop locations may receive more light than valley locations, yet may actually be cooler during the day due to dramatic exposure to the wind. However, at night the cold air may settle into the valley, causing colder conditions there than on the hilltop.

The warmth of the soil greatly impacts the level of microbial activity and therefore the potential availability of nutrients for the growing crop. Denser soils are much slower to warm. The ability of a soil to warm is often related to its level of aeration. Well-aerated soils heat more quickly; growers have used

FIGURE 1.8. Crimson clover, interseeded in August, is growing amongst this fall brassica. Rye and vetch will be interseeded in September.

tillage to supply such aeration to warm soils in spring. Tilled soils also tend to cool quickly, however, and may overheat in summer with associated impacts on soil life. An untilled well-aggregated soil full of life also heats up quickly due to its aerated nature and microbial activity yet is also much more resistant to overheating. This consistency of temperature is a major benefit of no-till, biologically active soils. Soil life thrives under these conditions and therefore provides steady, extensive nutrient availability to growing crops.

Not surprisingly, darker soils are capable of collecting more warmth than lighter-colored ones. The addition of organic matter in the form of compost and dark mulches often makes soils darker due to the steady decomposition of these materials and their recombining into humic substances and other stable carbon-rich materials. Very dark fertilizer materials can also aid in darkening soil: Examples include charcoal, humates, and bone-char (ground charred bones). They can be effective when top-dressed upon the soil. Growers sometimes plant crops through a black plastic soil covering in order to provide more soil heating, but this may inhibit the soil breathing process described earlier in this chapter. As sunlight contacts the soil surface, the air in the soil warms and expands, which forces some of the air out of the soil. Then as the soils cool from the lack of sunlight, a vacuum develops, which draws air back into the soil. Thus the interaction of sunlight and warmth are highly interrelated to the soil breathing efficiencies and therefore the ability of soil life to function.

The soil life itself generates a significant amount of warmth as it consumes food and releases energy. A highly active soil is therefore able to stay warmer during cool conditions than a soil with a low level of biology. This can often be observed during cold periods. This energy release and warmth allow soils to stay highly functioning even when covered by organic matter mulches, because there is biological activity in the mulch layer as well. Mulching a soil with organic material that is not highly biologically active can have the opposite effect, keeping soils cool in spring and delaying crop growth. The cooling effect of organic mulches on soils during summer heat, however, is generally beneficial to all soils regardless of tillage and structure characteristics.

Human Influence

Although management of the four elements is of primary importance, often the conditions of these elements are somewhat beyond the grower's influence. Nature has, to some degree, its own plan, whether that be a drought, a hurricane, a volcanic eruption that blots out the sun with particulates, or the long-term climatic conditions that led to a soil's formation. Humans have recently come to influence these elemental states on a large scale as well. Many of these human influences on natural systems are detrimental to the ability of the planet to sustain life. Humans certainly have the capacity to support the earth's ability to sustain life, and the benefit of humans to natural systems can be extensive. However, the impacts of human-caused pollution of agricultural systems have become widespread and very significant. This impact is likely to continue to intensify in the future. As growers, we need to understand and boldly face these impacts so we can adjust our agricultural systems and remain viable. Pollution has severely impacted all four natural elements: sunlight, air, water, and soil.

The sunlight that is the primary source for the energy giving us life has been increasingly blocked by particulate matter in the atmosphere as a result of human activity. The fumes of industry and the burning of long-buried, carbon-rich materials as fuels have emitted large volumes of volatile substances into the air. Herbicides applied to thousands of square miles of cropland vaporize into the air and are distributed across the earth. The nitrogen gas N_2, which makes up the greater portion of the air, is split in nitrogen production factories to create incredible quantities of liquid ammonias and solidified nitrogen salts. These nitrogen compounds are utilized in tremendous volumes by industry, in war, and as fertilizer. They readily volatize back into the air as various oxides of nitrogen, and the overall result is a highly disturbed nitrogen cycle. Radioactive substances are carried on the winds from nuclear explosions, accidents, and releases. Human created electromagnetic forces (EMFs) have greatly increased due to everything from electrical generation and transfer to radio, television, cellular communication, and electromagnetic warfare systems. All of these energetic disturbances are having a detrimental impact on living beings over the entire surface of the earth.

The waters of the earth are highly polluted as well. In particular the rain is contaminated, as are many sources of irrigation water. In the pre-industrial era, airborne water vapor was naturally distilled by evaporation and the primary dust in the lower atmosphere was salt dust lifted from seawater evaporation, which resulted in clean rainfall bolstered with natural trace

FIGURE 1.9. Chemtrails are a constant condition in our region, almost a daily occurrence. This suddenly began in the fall of 2016. The airborne particulate matter from them causes frequent white skies, as well as moon and sun bows like this.

elements. Now, however, the primary dust is pollutants such as half-burned carbon fuels and other "additives" from jet aircraft. This rain then falls through a disturbed gaseous environment in the air with the resultant acidification of the rain, often also containing the residues of vaporized herbicides and other chemicals.

Weather control and modification have been actively utilized by government and industry for many decades. Cloud seeding for moisture grew out of World War II smoke screen research and developed into a highly effective means of causing moisture to fall in specific regions. Many countries, including the US, are widely reported to be presently using this technology. However, causing rain in one region inevitably causes drier conditions in another. This has created a worldwide struggle for moisture control in the environment. Though cloud seeding for moisture is a relatively overt activity, most weather modification is performed in secrecy by the various militaries of the world. Weather control is a very important area for military concerns, and its usefulness in warfare is obvious. Though there was much outcry against the use of

FIGURE 1.10. Many trees in our region turn brown and lose their leaves very early. This photo was taken in August. Note the sparse canopies and jet chemtrail.

weather control in warfare after its extent was revealed during the Vietnam War, including an international treaty to limit its use in warfare, little to nothing has been done to contain it. At this time weather modification is no doubt highly evolved, and we are now in a time of extensive climate engineering. This being the end result of the human condition at this time, it is an activity brought forth by fear, greed, and foolishness, and as such is unlikely to be of benefit. Growers will probably benefit from knowing that weather conditions are highly manipulated, and specific conditions that occur in regions may be better understood in this context. Soils themselves are often treated with agricultural chemicals and overaggressive tillage, which, combined with all the factors above, results in a weakly functioning soil system. The living portion of the soil in particular is changed and disturbed. This results in poor agricultural production, dying forests, and collapsing ecosystems. The end result is desertification.

These disturbances to the elemental forces are particularly important to keep in mind when considering agricultural activities. They have profound

FIGURE 1.11. Twenty years ago spring-flowering brassicas like these were covered in pollinating insects; now we see only a handful. The decline of insects on this crop is greater than 99 percent. Many other insect populations have plummeted as well.

implications on how to fertilize, irrigate, till, make crop spacing decisions, and deal with pestilence, all of which have dramatic implications for profitability. The degradation of natural vitality also affects humans. Illness increasingly reduces the potential of the human both on a physical and mental level. This makes well-grown vegetables all the more important, as these vegetables are a critical component of a diet for health.

For the grower to be successful at this time of profound environmental disturbance, an open-minded approach—one that emphasizes flexibility in production systems—is critical. The growing environment is rapidly changing, and the ability to change with it is primary. As growers we must continue to be able to observe these changes and assess their impact on our farms in order to be successful with our growing efforts.

CHAPTER 2

Balancing Crop Growth

The forces of nature are the primary source of the energy required for plant growth. How these forces are influenced by growers' actions and the materials (such as fertilizer) that growers apply has direct ramifications on the energy available for crop production. Thus, observing crop growth and developing an understanding of the influence of nature's energetic forces on crop growth is a critical skill for growers, especially for growers who wish to reliably produce abundant crops without the use of pesticides. Growers develop their proficiency through careful observation of the natural world over an extended period of time. The maintenance of an open-minded approach with a deeply held desire to learn is critical. The study of published material on the natural world is also of assistance. Biodynamic literature is particularly insightful on this subject. Crop growth does not occur from physical materials alone but from the energetic reactions among them. Consider, for example, the basic process of photosynthesis: It is the energy from the sun that provides for the combination of carbon dioxide and water to produce sugar. As growers we work with energy and forces constantly, and it is well within our abilities to do so. Indeed, failing to understand how the energies and forces of the materials we use impact the growth and development of plants may lead to unnecessary crop failures.

Picturing Balanced Growth

With refined observation and practice, it becomes easy to identify balanced, healthy crop growth. Many physical characteristics are reliable indicators. The seedlings spring out of the ground quickly. The initial leaves or cotyledons are relatively large and remain green for an extended period of time. The growth rate is steady and uninterrupted from germination through to seed-bearing maturity. Maturity is not delayed. The plants are sturdy but not brittle; they flex when they are brushed against or harvested. They are capable of standing upright during heavy winds and rain without lodging. The stems are thick and have balanced internodal distances. The leaves have a consistent green coloration (or if their pigmentation is other than green, those colors are extensive and vibrant) that is not too light nor too dark. There are no signs of mottling; the leaf edges are not scorched, and the leaves do not twist or roll. Leaves have tangible thickness and are not excessively hairy. The leaves are large and plentiful and appropriately positioned for maximum collection of solar energy.

Freshly dug roots are vibrant white, particularly at their tips, with no discoloration. There is an appropriate balance between the number of horizontal feeder roots and taproot size. Roots do not appear excessively hairy. Soil life around the roots is abundant, particularly earthworms. The soil sticks to the feeder roots, often encasing each separate root. If pigmentation is appropriate in the roots of the crop, that pigmentation is extensive and vibrant. The amount of root growth and top growth is balanced; neither is excessive, both are in proper ratio.

From flower to fruit to seed, the quantity is abundant yet not excessive. There is colorful pigmentation and vibrancy of flowers and fruits. The flowers all fill to fruits, and the fruits are carried to full maturity. The fruits are completely filled out, large, and heavy. They evenly ripen. The seeds are plump and heavy and mature in a timely manner. The seed is of high germination percentage and quickly germinates. The vegetables are very well flavored with a relatively high sugar content. Customers, including small children, readily consume them. Vegetables are neither hard and fibrous nor soft and watery. If appropriate, they develop pungencies and odors that are obvious and pleasing. They keep very well post-harvest, often dehydrating over time if left in storage rather than rotting. The benefit to consumers' health of consuming this food is obvious.

All of these conditions go hand in hand with insect and disease resistance as well as high yields. Most of these conditions also all happen at once. In other words, when all is well, all these conditions are present. This leads to ease of harvest, customer appreciation and enthusiasm, lengthy sales periods, and all the other benefits of a well-raised crop, not the least of which is the opportunity for profit. Often, however, growing conditions may not be tending toward this picture of balanced health, and in these cases growers may need to make some adjustments arising from a thorough understanding of those forces that influence and create balanced conditions.

Crop Plant Polarities

In the physical world, forces work in polarities. Examples include positive and negative, warm and cold, day and night. The yin and yang symbolize these polarities particularly well.

These polarities are at work in plant development, and they can be described as the force of growth and the force of reproduction. They can also be compared to the forces that form interiors and exteriors, or to gravity and levity, or to the female/male polarity. In terms of crop development, certain materials and forces aid in growth or sizing of the plant, and others induce flowering, fruit, and seed production. As in the principle of yin and yang, there is some intermingling of forces in the polarities (see figure 2.1). For example there are always varying degrees of female qualities in male entities and vice versa.

The actions of a grower are a primary influence on the balance of forces, with the capacity to increase or decrease the strength of the force of both or just one side of the growth/reproduction polarity to varying degrees. Because of this, it is fundamentally important for growers to understand their actions and conduct them appropriately in order to assist crops in their expansion to an appropriate size and as well to bring them into a condition of abundant fruiting and seed. Appropriate action can bring a cabbage head to a large size and withhold it from flowering prematurely, for example, or bring a sweet pepper to ripeness before frost threatens. To guide crops in this way is well within a grower's ability, and not to understand actions that influence this balance risks unnecessary failures. Perhaps the best way to understand these polarities is to examine conditions taken to the extreme.

Excessive growth forces can lead plants to enlarge quickly, yet the vegetation is notoriously weak. This condition is often referred to as rank growth. This weak growth is prone to collapse, as in lodging, and injures easily from stresses such as wind, rain, frost, or physical contact. The excessively lush crop grows over itself and often starves some of its vegetation of sunlight. The harvested vegetable portion has a soft texture, and quickly rots upon storage. The flavor is generally described as watery. The pigments are lacking and the coloration is dull or washed out. Taproots are limited in size, with excessive small horizontal feeder roots, which give them a hairy appearance. Plants have "too much top"—the growth of the top or aboveground portion of the plant far exceeds the growth of the roots.

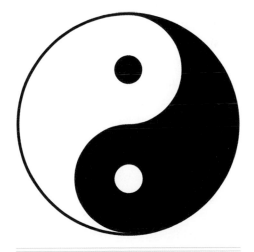

FIGURE 2.1. The principle of yin and yang is an ancient and accurate depiction of the polarity of forces and interaction. The circles within the areas of opposite coloration represent the lack of absolute purity of forces within the polarity.

FIGURE 2.2. Our garlic shows more intense pigmentation each year since we converted to no-till. When we were tilling the soil, this variety of garlic was always plain white.

FIGURE 2.3. The radish, *left*, shows the effect of strong growth force. The plant, *right*, which is the same variety, shows strong reproductive force.

Crop maturity is delayed. Flowering occurs later than usual, the fruits often few in number. Seed is set poorly and ripens very slowly. There may be classical signs of deficiency of the nutrients associated with flowering such as phosphorus, sulfur, magnesium, manganese, and silicon.

Crops that are under the influence of excessive growth force conditions are often attacked by insects and disease. This condition is common in vegetable crops because growers are prone to the overuse of materials that bring on sizing. (Growers are much less likely to overuse materials that constrict growth.) To our benefit, nature sends in disease and insects to destroy these overgrown, imbalanced crops, as they are unfit for consumption by humans. However, growers often strive to defend them by applying pesticides. This then adds chemical toxicity to a crop so inherently imbalanced that it brings disease and pestilence. With so many crops being produced under these conditions in today's commercial agricultural systems, human health has greatly declined.

A common indicator of excessive growth force in vegetables is the aphid, often linked to excessive nitrogen fertilization. Another common difficulty is slugs, which are particularly attracted to the soft, watery nature of these crops. Indeed, a whole host of insects may be attracted to these crops. With their skewed metabolism due to these excesses, the plants are unable to form the complex compounds that are the physiological basis of resistance to disease and insect pests. Under these conditions fungal diseases are particularly prevalent as well.

Excessive reproductive force often has an impact opposite to that of the growth force. Growth is generally stunted and yields of harvestable portions very low. The aboveground portion or top is particularly small by ratio to the root growth; the roots may be of relatively small size as well. There is stronger development of the taproot than the feeder roots. The entire plant is hard and fibrous, with strong, pungent off-flavor and dry texture. Thus, these crops are particularly unattractive to customers. With excessive reproductive force, flowering response of the plants comes on very early. Blossoms are profuse, yet fruits and seeds are often small due to insufficient growth force. Yields invariably are low due to small size and inappropriately early flowering, when plants have insufficient top growth to support adequate fruit and seed maturation. The leaves may be markedly pigmented, but coloration is often inappropriate. There are often signs of deficiency of one or more nutrients associated with sizing such as nitrogen (N), potassium (K), or calcium (Ca).

Influencing the Polarities

With this understanding of balance versus imbalance in vegetable crops, the next step is to examine how growers can affect the relationship between growth force and reproductive force. Obviously, applying fertilizer is one way to influence plant growth, but growers do well to look first at conditions of the primary elements: soil, air, water, and sunlight. The volume and intensity of the combination of these factors influence the imbalance or balance of the growth/ reproduction polarity, as shown in table 2.1, which summarizes the general environment and fertilizer influences on the polarities.

The polarity between some of these factors is tangible, such as wet versus dry, light versus shade, and valley versus mountain. All of the growth factors are highly interconnected, just as all of the reproductive factors are interconnected. In addition, each factor is connected with its polar counterpart. How these factors combine in field conditions strongly influences whether a crop is in balance or is skewed to one polarity. Let's have a look at the various factors in terms of vegetable crops.

WATER

Water has a considerable influence in its ability to increase the force of growth. Thus many growers add water frequently to their crops in order to size them. However, many growers also recognize water's ability to influence and potentially increase disease in crops. With these two points in mind, it's clear that

TABLE 2.1. Environmental and Fertilizer Influences on Growth and Reproductive Forces of Crop Plants

Growth Force	Flowering and Reproduction Force
Water/wetness/stillness	Air/dryness/wind
Moon	Sun
Night	Day
Shade	Light
Cold	Warmth
Valley	Hilltop
Clay	Sand
Organic soils	Mineral soils
Earthly/Biodynamic preparation 500	Cosmic/Biodynamic preparation 501
Interior	Exterior
High plant populations	Low plant populations
Loose seedbed (tillage)	Firm seedbed (compaction)
Alkaline soils	Acid soils
Bacterial dominance	Fungal dominance
Nitrogen	Carbon
Nitrate nitrogen	Ammonium nitrogen
Calcium	Silicon, magnesium
Potassium, sodium	Phosphorus, sulfur, many micronutrients

irrigation is one of the most impactful influences on crop balance that growers commonly utilize. Irrigation or watering can be very beneficial, yet can also easily create rank growth conditions, throw soil microbiology out of balance, and, through disruption of the soil air:water ratio, lead to a "watery" condition in crops. When making decisions about whether and how much to irrigate, it is important to think through to what extent crops will remain wet. How long will the leaves be wet with dew or irrigation water over the course of the day? This depends not only on how much irrigation water is applied but also on soil moisture, relative humidity level, presence or absence of breezes, and the possibility of rain in the near future. Weighing all of these factors can lead to a more informed decision regarding the influence of water on crop balance.

AIR

Air is the great balancer of the watery influence. Air (wind in particular) dries out crops and keeps them from becoming rank. Excessive influence of wind and air can easily stunt crops, though, because plants will harden and "hunker down" in windy conditions to avoid physical damage. Growers manage air conditions not only by considering where fields are located in terms of wind exposure but also by crop row placement, density of planting, use of raised beds or hilling, and setting up physical wind breaks. For example, if a site is generally under a watery influence, it may be best to space out crops more widely; for sites under an airy condition, it is often best to tuck them in closer.

LIGHT

The night/day length also has tremendous ramifications for the growth and flowering of plants. Many plants' primary trigger to invoke flowering is photoperiod (day length or night length). In general plants respond to a seasonal time of particular day/night length that is a powerful signal to reproduce. It is actually the length of the dark period that is the signal to begin flowering and reproduction for many vegetable crops. This condition is often controlled in indoor environments through the use of grow lights and extension of dark periods. However, field growers can simply pay attention to a crop's known reproductive season and plan accordingly. For instance, in order to have a consistent supply for market, crops like cilantro and arugula need to be planted in closely timed successions in spring because they will bolt to flower quickly then. In late summer these crops do not rapidly go to flower, so the successions can be spaced farther apart in time.

The two solstices have a pronounced influence; plants sense the diminishing darkness after the winter solstice and the increasing darkness following summer solstice. The nature of the response differs among crops; for instance, a radish plant generally bolts only during spring's decreasing darkness, whether it was seeded in the fall or in the spring. A carrot plant will generally not bolt unless it has experienced the increasing darkness of fall, which signals the plant to flower the following spring. This means that carrot seedlings emerging from the soil after the winter solstice usually do not bolt in spring, but those germinating before the winter solstice often do, especially as seeding date is moved back from the solstice by a month or so. Of equal importance is the great diminishment of flowering of winter-hardy crops after the summer solstice. This period allows for extended growth and harvest of many greens and root vegetables, which will not bolt until the following spring.

When considering the influence of shade and sunlight on crop growth, it's a common misconception that exposure to direct sunlight is the factor that makes a plant gain size. Plants do require sunlight for photosynthesis and hence the energy for growth, but plants *grow*—or increase in size by cell division—at night. The direct force of sunlight, especially intense sunlight, generally causes plants to stay smaller and more constricted than they would be if exposed to limited light. Shade-grown plants put on height rapidly in an effort to reach up toward the sun as well as grow large leaves to help collect what sunlight is available. Growers use various means to limit sunlight in efforts to increase growth, such as covering beds with shade cloth or row covers, as well as planting short crops in the shade of tall crops. Consider the impact of a white fabric row cover. The cover reflects the sunlight while keeping in moisture and reducing airflow. The net effect is increased growth, because the three factors all work together to increase this influence. Of course if overdone, or combined with the strong influence of other growth-enhancing factors, this can cause the plants to overgrow, creating lush conditions that are prone to disease and insects like aphids.

The moon also has dramatic impact on crop growth, particularly at the full moon, when the sunlight it reflects creates a partial light similar to that of a shady spot in daytime. This level of light further stimulates plants to expand. The full moon also influences the earth's water cycle, which often brings moisture during its period, and this also makes the full moon a time of increased growth. This is a benefit if this is what a crop needs at that time, but a detriment if a crop is already growing in a condition of excessive growth force. When the moon is full, the moon and sun are on opposite sides of the earth. When the moon is moving with the sun across the sky during the new-moon period, these influences are not in place and therefore there is greater reproductive force.

HEAT

Warmth and cold are fully integrated with the aforementioned influences. Cold develops under wet conditions lacking in sunlight; warmth develops with dry, sunny conditions. A soil with much air and little water in the pore space warms more quickly than a saturated soil, for instance. Warmth is also highly related to a plant's flowering response, second only to photoperiod. In some crops, warmth may even be a dominant factor, because some crops require a period of alternating warmth and cold to bring on flowering. Under cold conditions plants generally withhold their flowering, because this often coincides with a lack of other natural factors for successful fruition, such as insect pollinators.

Managing for Balance

Grower actions can significantly shift the balance of the growing environment of a particular site, of individual crops, and of the soil.

ENVIRONMENT

In a valley location the environment is often watery, still, shady, and cool, whereas a hilltop often offers airy, dry, windy, sunny, and relatively warm conditions. There are exceptions, of course, but the important consideration is which conditions are dominant in each of the crop environments a grower is managing. Some locations are more balanced, some more extreme. In a more extreme environment, appropriate actions by the grower will be necessary to establish an appropriately balanced growing area. Consider the valley and the hilltop, for instance. Perhaps trees need to be removed in the valley location to allow in more sunlight, warmth, and air. Maybe trees need to be planted on the hilltop location to buffer the wind and increase moisture, as well as provide some shade.

Soil characteristics vary among locations as well. Valley locations often have high clay and organic matter content. Hilltops often have a sandier, more mineral soil. This is due to the hilltops being prone to erosion, which moves the finer clay particles and lighter organic matter downhill, leaving the coarser particles behind. Once accumulated in a valley location, the organic matter and clay materials support more expansive growth, which leads to even greater levels of organic matter in the soils.

Natural forces thus can shape conditions such that a site does not offer a balanced polarity between growth and reproductive forces. This provides for diversity in the landscape, as nature has adjusted with development of specific soil biology and plants that grow appropriately in each unique environment. By studying the plants that persist naturally in a location, we can infer the kind of crop that is best suited to that growing environment. We can also better prepare to adjust conditions for a crop that would not naturally benefit from such an environment.

When Biodynamic practitioners discuss this overall polarity of a site, they often use the terms *earthly* and *cosmic*. The earthly force is akin to gravity. The cosmic force moves away from the earth toward the cosmos—it is akin to the opposite of gravity, which is levity. The earthly force keeps the plant close to earth by increasing leaf sizing and delaying flowering. The cosmic force encourages the plant upward as the flower stalks reach toward the cosmos and in particular the sun. Flowers reflect the cosmic beauty of color derived from the interaction with sunlight. Seeds are meant to spread out, with some kinds being light enough to float up and away from the terrestrial environment. Biodynamic growers typically utilize Preparation 500 (horn manure) to increase the influence of the earth polarity and Preparation 501 (horn silica) to increase the influence of the cosmic polarity. The preparations are based upon Rudolf Steiner's recommendations for agriculture. In this case they are either manure or ground silica buried in a cow horn during certain times of year. These are then dug up, stirred into water, and applied in a very dilute quantity to the growing area.

PLANTS

The cosmic (reproductive) force is concentrated on the exteriors of plant organs, while the earthly (growth) force is concentrated in the interiors. The

FIGURE 2.4. These leafy crops have a radiance due to their strong exterior, which also helps to keep them free from insect damage.

hulls, shells, bark, skin, and feathers of living organisms contain the materials of exterior force that hold in the organism. These surfaces are directly exposed to the cosmic. The interior starches, proteins, flesh, and bones contain the materials of the earthly force of growth. Examining crops with an eye to the ratio of interior to exterior volumes is useful in determining its present state of balance, and thus provides information on how to guide that crop. The volume of the exterior material is generally the amount of hull, shell, skin or other outer casing; the interior volume is what is contained inside this material. Here it may be useful to picture the ratio of external and internal substance of a large corn kernel to that of a very small seed such as amaranth, quinoa, or teff. In general growers seek to encourage greater crop size, and therefore many crops end up excessively swollen with weak exteriors—an imbalance of force and material. This common practice of pushing crops toward excessive growth force, and thus the swelling of the interior, has also unfortunately assisted the excessive interior growth in humans through the consumption of these crops. The manifestation of imbalance can be seen in their overweight nature, rampant excessive cellular growth (cancer), clogged arteries, and overactive nervous systems, as well as their lack of skin tone, elasticity, and radiance (cosmic force).

As the forces condense into matter, plants take on the dominant nature of these influences in their physical makeup. Thus when we gather materials that grew under a certain influence and recycle them into our growing systems, they bring that influence into the growing system. Even something as simple as the selection of hay for mulch can influence the balance of polarities. For example, you can choose the choice of an early-cut spring grass hay for a surface mulch fertilizer rather than a more mature hay—say, a first-cutting of hay from September that includes large amounts of exterior coatings such as dried hollowed stems and shattered seed heads. If we choose to apply materials to the growing area that are young and lush, we see a lushness in the resulting growth of the crop. Materials that are more mature and hardened bring a hardness into our crops.

Planting density also dramatically impacts how the forces can influence a crop. Under conditions of close spacing, the crop canopy retains water longer, air does not freely circulate, and the plants shade one another from the sun. These conditions aid plants in their growth to some degree but result in rankness if taken too far. On the other hand low plant density brings in more air and light and dries crops more quickly. This results in much sturdier growth, yet overexposed plants will not grow as quickly. As growers gain knowledge of their basic field conditions, they develop greater ability to select appropriate planting densities. In a highly exposed field, it is often useful to space plants more tightly than usual. In a moist, still valley location, it is often better to keep the plants farther spread to assist with balance.

SOILS

The choice of tillage, reduced tillage, or no-till management also has significant impact on balance. Tillage is inherently destructive to a segment of the soil life, particularly the fungi, and results in a quick release of growth-enhancing nutrients from the destroyed organisms as they decay. Often the bacteria come to dominate this environment of fast decomposition, and the net result of this nutrient release is a rank lushness in the crop. Fungal organisms, by contrast, are more primary in the release of the hardening/ reproductive materials from the mineral portion of the soil and bringing them into forms available for crop uptake. The dominance of the bacteria resulting from the physical impact of tillage also often causes the development of more alkaline soil conditions, which is another indicator of the potential for excessive growth force. The destructuring of the soil from tillage sets up conditions for salt buildup, particularly in dry conditions like arid regions and greenhouses. This concentration of salts also leads to alkaline soil conditions. Acid soil conditions often develop in high-rainfall regions. There fungal organisms are prevalent and assist in balancing growth in the watery environment. A pH level (a measurement of the ratio of H+ ions to hydroxide ions, OH–, in solutions) of 7.0 is chemically considered balanced. However, most vegetable crops grow well at a soil pH of about 6.5, and it may be best for growers to consider 6.5 as the point of balance. As soil pH approaches 7.0, there is a tendency toward the force of growth. On the other hand the impact of extreme soil acidity is stunted plant growth, as in excessive reproductive force.

No-till brings the capacity for greater interaction of crops with mycorrhizal fungi and a potential for a more fungally dominated soil. These soils often have slightly more acidic conditions. In the firm seedbeds of a no-till system, there is no quick release of soluble nutrients from tilled-to-death organisms. Instead the untilled soils with their fungal networks have a greater likelihood of biological balance, and they offer slow, steady release of mineral materials that are often much more difficult to access in a tilled soil. These minerals include nutritive elements like phosphorus and magnesium as well as many of the micro and trace elements. These minerals greatly aid in the hardening of the crop.

Balancing Through Fertilization

Fertilization can have dramatic influence on the balanced state of crops. Nutrient addition through fertilization can guide a crop to desired conditions in terms of growth and reproduction, but fertilization may result in imbalances or undesired outcomes if inappropriate materials are utilized. Likely the most important nutrient ratio to pay attention to in crop production is the carbon to nitrogen ratio. The ratio of the abundance of these nutrients in soils, fertilizers, mulches, crop residues, or composts is referred to as the C:N ratio. It is a ratio over which the grower has much influence. Nitrogen-rich materials are of course the primary materials used to increase growth, while carbon-rich materials rein in this force and harden and build the structures of fruition. Many growers are familiar with the importance of C:N ratios in terms of composting. It benefits a grower to expand this understanding to field production as well, because manipulating the C:N ratio in the field can greatly aid in rebalancing growing conditions.

Nitrogen-rich materials are concentrated in lush young foliage as well as the interiors of organisms. Nitrogen's release for plant uptake is also quickly brought on by tillage and subsequent bacterial activity. (It is helpful to remember that protein is a common source of nitrogen in an organic form.) As well, nitrogen is concentrated in watery valley conditions. Legumes in particular concentrate nitrogen from the air into the soil along with themselves through their association with rhizobial bacteria. When nitrogen needs to be increased the addition of materials that are made of, or have otherwise grown under these conditions of concentration, can be utilized.

Carbon-rich materials are often the dried-up, mature plant parts or exterior parts of organisms. These materials tend to be brown in coloration. Fallen leaves, straws, barks, wood chips, and seed hulls, as well as the sugars produced by photosynthesis, are all high in carbon content. In order to be reassimilated into living systems, mature high-carbon materials like lignin and cellulose may require fungal decomposition as opposed to the speedy bacterial decomposition of nitrogen-rich materials. When rank growth needs to be reined in, one of the most basic management approaches is to increase the carbon inputs into the system. In the case of high nitrogen levels, the increase of carbon levels to match in an appropriate ratio is key to abundant, healthful growth.

Management of the C:N ratio in the field is accomplished through grower activity (such as planting specific cover crops), as well as by applying fertilizers, mulches, and compost. The selection of which cover crop species to grow is important, as well as decisions about when to terminate them. Grasses are often carbon-rich, whereas legumes are often nitrogen-rich. Both are higher in C:N ratio when young, and have a lower C:N ratio when mature. The choice of mulch materials has an influence, and the management and composition of crop residues can also be significant. The nature of the crop that preceded the planted crop also needs to be considered as it will have influenced the C:N ratio. Management of soil air and soil biology also directly impacts nitrogen levels because nitrogen, as a relatively inert gas, is abundant in the air and can be brought into living systems by the activity of soil microbiology. Environmental factors can directly impact the C:N ratio as well, such as when a heavy rain leaches nitrogen from the soil.

Carbon intake by the plant is largely in the form of carbon dioxide gas released from decomposition in the breathing soil, often from directly beneath the plant canopy. Materials and microbes in soil influence this release. The carbon is then reintroduced into the soil system by root exudates or carbon-rich substances, as well as the eventual decomposition of the plant. How soils are handled in terms of tillage and air:water balance also directly influences C:N ratios and levels because tillage and irrigation influence soil biology and the ability of the soil to breathe.

To understand the influence of nitrogen fertilization, it is useful to take a deeper look at nitrogen forms in plants and soils. The nitrate (NO_3) form of nitrogen is the primary nutrient associated with the growth force in plants. Ammonium (NH_4) is an associated but often less abundant form; it enhances flowering and reproduction. Soil and environmental conditions largely determine the dominant form of nitrogen in a particular soil. Interestingly, the conditions that lead to an accumulation of the growth force also lead to an accumulation of nitrate, and the factors that contribute to the flowering response help keep nitrogen in an ammoniacal form. Thus, increasing ammonium in the soil is often not easily accomplished simply by adding an ammonium-based fertilizer. If the soil is dominated by the growth force, that ammonium will be quickly transformed into nitrate. Such conversions occur with many other nutrients as well. When a soil is low in a given nutrient, the lack may be due to greater environmental or biological conditions. Adding a fertilizer containing that nutrient may be ineffective if the underlying conditions are not addressed.

Many nutritive elements used as fertilizer are capable of influencing the strength of the force of growth or reproduction. In Biodynamic agriculture the polarity is often described as calcium/silica polarity. Calcium is the growth side and silica is the reproductive side. Calcium is concentrated in the interior structures such as bones; silica is concentrated in exterior structures including skin, hulls, and shells. In soil testing as well as in analysis of human health and supplement recommendations, the calcium to magnesium ratio is often examined. Nutrient ratios are certainly worth tracking through soil and tissue testing as well as through physical indicators, and considering when making fertilizer decisions. The ratios of other growth nutrients (potassium and sodium) to the reproductive nutrients (phosphorus, sulfur, manganese, zinc, copper, chloride, boron, cobalt, and iron) as well as to one another are often aided by soil and tissue analysis (chapter 8 has details on such analysis).

Nutritive elements are best kept not only in balance with one another but also in relatively high levels of availability for crop growth. Soil and tissue analysis can be useful in developing this strong and balanced nutrient relationship. The addition of fertilizer materials aligned with a specific polarity often enhances that polarity and can be utilized to direct a crop toward balance or slightly away from it. Examples of potent growth mineral fertilizers are calcium nitrate, sodium nitrate, and hydrated lime (calcium hydroxide). These materials have constituents that all enhance growth. Examples of plant reproductive/flowering mineral fertilizers are talc (magnesium silicate), manganese sulfate, elemental sulfur, and Epsom salt (magnesium sulfate). More balanced fertilizers contain nutritive elements from both polarities such as calcium silicate, rock phosphate (calcium phosphate), potassium sulfate, or gypsum (calcium sulfate).

As stated earlier, many nutrient availability conditions are influenced by the environmental and biological condition of the field. Thus balancing of the various ratios of nutrients is tied to soil management techniques as well as fertilization. It is best to keep nutrients within a polarity, such as the calcium: potassium ratio, in balance (refer to table 2.1). Balance across the polarities, such as the calcium/silica polarity mentioned above, is also important. For balanced growth overall, growers are seeking balance among all the nutritive elements. In general an excess of any single nutrient causes imbalance both in relation to other nutrient elements of the polarity as well as to nutrients that are in the opposite polarity. There are circumstances in which increasing

TABLE 2.2. Nutrient and Water Needs During Different Growth Periods

Growth Period	N	C	C:N ratio	H$_2$O	P	K	Ca
Leaf period	high	low	low	high	low	low	low
Flowering period	low	moderate	moderate	moderate	high	moderate	moderate
Fruiting period	low	high	high	low	moderate	moderate	high

a certain nutrient to be better balanced can significantly impact the availability and balance of other nutrients. Gentle efforts toward balance and increase while avoiding excess is a beneficial approach, but often easier said than done. With the current dramatic imbalance of nutrient levels of plants and animals, this may require careful observation, tracking, testing, and guidance. However, it is important to remember that nature is also seeking balance and can assist us in our efforts.

Once general balance is understood, and perhaps has been practiced and achieved in various crops and conditions, the next step is for growers to manipulate these conditions away from balance to some degree when doing so assists in achieving specific crop production goals. This includes managing conditions (such as water status) as well as using fertilizers. The goal might be to stimulate more leaf growth, for example, or to bring crops to flowering and fruiting earlier than usual. Table 2.2 summarizes general nutrient management and fertilization through the three growth periods of crops; this table is adapted from Korean Natural Farming principles.

Table 2.2 shows the conditions that lead to extensive leaf growth: high nitrogen, low carbon (low C:N ratio), with much water. All this moves a crop toward expansive growth conditions. To move the crop toward the flowering stage, reduce nitrogen dramatically and increase carbon (shift from a high to a moderate C:N ratio). Cut back on watering, increase phosphorus dramatically, and start to increase calcium and potassium as well. For the fruiting period the use of materials that encourage the reproductive force continues with nitrogen still low and carbon increased for a high C:N ratio. Water is very reduced, phosphorus lowers to a moderate volume, potassium remains moderate, yet calcium is increased in order to assist fruit sizing. These conditions can be achieved through various management influences combined with side-dressing and foliar fertilization.

Applying the Concepts

With this concept of plant growth periods in mind, let's look at how the influences of basic field environment and the grower's actions and fertilizer management affect a common vegetable scenario, the fruition of tomatoes in August. In the Northeast in August, the sun is diminishing, the air is often still, and humidity and rain are increasing, particularly around the full moon. All of these conditions lead to excesses of the growth polarity and thus to fungal disease. If the field is on a hilltop and crops are well spaced, better balance may be achieved, conditions will remain balanced, and fungal disease will likely not be severe. But if the field is a valley location already high in nitrogen and potassium, which both favor rank growth, trouble is definitely brewing. Hopefully in this case the grower in the valley location has already applied high-carbon materials to dilute the nitrogen, mycorrhizal fungal relationships are in place, and trace minerals are being supplied. It may be wise to apply calcium materials to assist in balancing the potassium and nitrogen on the growth polarity as well as to enhance fruit sizing. Reproductive materials such as horsetail tea (to provide silica), seawater extract (a source of magnesium chloride), and vinegar (acetic acid) applied as a foliar spray or side-dressing would further harden the crop and protect against rankness and fungal disease.

Another example is a hilltop crop of cabbage. In this environment it may be best to plant the cabbages with a tight spacing and provide a steady water supply and a fertilizer rich in nitrogen, potassium, and calcium, which are nutrients that will move this crop toward large head size and delayed bolting.

These examples demonstrate how nitrogen-based fertilizer can be useful under one set of conditions but hazardous under another. They also suggest that certain natural environmental conditions are best suited to various crops, and growers do well to understand which crops naturally grow well in each particular environment (field) they manage. However, growers often wish to plant a diverse mixture of crops all in one field. In this case it is likely that some of the crops will be challenged in the given environment. These crops often are best approached more cautiously, until the methods and means of proper adjustment can be discerned. So it may be most profitable for growers to plant a majority of crops that are known to grow well in the given location. Planting a smaller proportion of crops that do not naturally thrive offers the potential for growers to practice their management skills until appropriate techniques are developed. These challenging crops are important for growers to undertake to develop their abilities, but they cannot be relied upon for profitability.

Monitoring the relative state of balance of individual crops is important, but so also is monitoring the influence of growing practices on overall field conditions. Over time, as your understanding grows, actions can be taken to alter the conditions of air, water, sunlight, and soil toward a more defined goal. The best measurement of progress often comes from a grower's own field observations, though some soil and tissue analysis is also useful. Obvious signs of progress are diminished insect and disease difficulties. Significant improvement in response to a grower's action is heartening, but even small changes can lead to insight into what is bolstering disease and insect resistance. Progress might manifest as root maggots no longer damaging the rutabagas, even though they are still damaging the turnips, for example. Maybe the following season there will be less turnip damage, and after a time even the turnips will be free of damage. A farm that has struggled with flea beetle damage on all cabbage family crops may see damage steadily reduced, with damage ceasing first on the *Brassica oleracea* species, then the *napus*, and perhaps eventually all crops are developing free of flea beetle damage. Adjustments may take time and steady grower action, so it is valuable for growers to carefully observe what is leading to improvement.

Sometimes it is easy to "read" the general overall appearance of crops in a field in terms of balance. Questions can be asked to guide the eye to notice significant details. Is the crop too rank? Is it too sparse and stunted? Are there obvious nutrient deficiency signs on the leaves? How are the yields? How are the coloration and pigmentation? And so on. To help with this assessment it can be useful to grow some crops that act as easy indicators. Pigmentation is an easily spotted indicator, and specific plantings observed over time can yield useful information. There are crops that develop their characteristic pigment only under relatively balanced conditions. Red Salad Bowl lettuce and Red Sails lettuce, among other lettuces, are excellent examples. Under conditions of excess growth forces, the leaves are quite green, but under balanced or slightly reproductive conditions, they are bright red. Red Salad Bowl, in fact, can approach the red pigmentation of the reddest of red lettuces. Under excessive reproductive force the crop may still show red pigment, but it is often yellowed and growth would be stunted. Another crop of use as an indicator is Purple Top White Globe turnip—how purple is the top? Watermelon radish—how red is the interior? As garlic bulbs are drying, are they white, pink, red, or violet? As balance is achieved and the strength of the forces increased, garlic pigmentation moves to the more intense reds and violets.

FIGURE 2.5. Red Salad Bowl lettuce is an excellent indicator plant because of its significant capacity for pigmentation. We have seen this lettuce develop almost entirely green under excessive growth force conditions, and even darker red than this bed of plants under more reproductive force conditions.

Another indicator crop is the small red radish. In this case the quick-growing radish can be used to give a grower a comparison of the size of the top growth to the size of the root. This yields insight into the balance in the growing environment. Other growth conditions can also be examined. For instance, if spring-sown, how fast did plants bolt? Were there any signs of nutrient deficiency on the leaves? How long did the cotyledons remain green? Was the radish root tough and fibrous, soft and watery, or just right? How about the red pigmentation? Insect or disease problems? These kinds of questions can be asked about all crops in a grower's fields as a way to continuously gain insight. Weeds can also be useful in this manner, as they are usually easy to find in place whenever assessment may be required. The species of weed that is dominant in a growing environment, as well as its relative state of growth, can be observed for indications.

CHAPTER 3

Preparing Land for No-Till

Converting the vegetation of an area of land from the more natural conditions of perennial plant coverage to that of selected annual crops is often a necessary part of starting a new vegetable farm or garden, or expanding an existing one. This process, and the continuing work of maintaining production areas from year to year, has often employed extensive tillage in order to fit the land into a state that is acceptable for seeding and planting. Tillage has become both excessively utilized and extreme in its damage to soil functions, especially with the development of more powerful equipment. Many growers are now seeking to limit this damage by being much more careful and judicious in their use of tillage equipment. This is often referred to as *reduced tillage*. When systems are developed that require essentially no disturbance of the soil, *no-till* has been achieved.

In terms of soil health, it is best to reduce tillage as much as possible, but conditions may dictate the need for occasional tillage, as in the case of initial conversion of an area to vegetable growing, or to incorporate soil amendments thoroughly into severely depleted soils, or to control particularly noxious perennial weeds. For purposes of this manual, no-till means that tillage is not utilized for seedbed preparation. However, soils may still be slightly disturbed at some times, such as when opening a furrow to set transplants in, hoeing to cut weed roots, or harvesting root vegetables.

Pros and Cons of Tillage

Tillage, though inherently detrimental, can also provide some benefits to the vegetable grower. The natural state of agriculturally suitable land is a cover of perennial vegetation such as grasses, shrubs, and trees. Tillage is the traditional method to convert such land to vegetable production. Most vegetables are fast-growing annuals, and annual plants are nature's response to disturbance.

Thus vegetables, to some degree, are a reasonable plant to cover a tilled soil. Tilling not only quickly converts land out of its natural perennial growth but also creates a surface that is appropriate for seeding or planting. Tillage destroys weeds and mixes fertilizers and organic materials into the soil profile and can break up plow pans and surface crusts. Thus tillage may have beneficial results in terms of air and water movement, soil temperature, and residue decay. The need for many of these improvements, however, may actually arise from inappropriate past tillage events. To some degree tillage may lead to conditions where more tillage is needed, a sort of tillage treadmill effect.

For growers to best utilize tillage for potential benefits and avoid perpetuating the need for tillage, they must first identify a clear purpose for tilling and understand the damage that tillage events may cause. This provides the best chance that they will achieve their goal and inflict the least possible damage. Though nature is forgiving, over time repeated disturbance by tillage can wear down a soil's ability to effectively recover. The detrimental impacts of tillage were detailed in chapters 1 and 2. The deterioration of soil structure leading to soil crusting and subsurface plow pans, increased erosion of soil, destruction of soil life, and the dramatic impact on the soil temperature and imbalance of soil air:water biology and nutrients are just the short list. The ramifications of these conditions on crop health can be extensive. Weeds respond to the disturbance with rampant growth. Damage and imbalance to the soil biology lead to nutrient imbalances that if unmanaged by growers quickly lead to disease and insect assault as well as poor growth with all manner of production difficulties, leading to lack of profitability. Fortunately, there is a better way.

The transition from tillage systems to reduced tillage and finally to no-till is often gradual. As well, no-till may not be a permanent field condition, and some form of tillage may be reintroduced in order to achieve a specific goal such as eliminating an infestation of aggressive perennial weeds, with an eventual return to no-till when that goal is achieved.

Growers whose systems presently incorporate tillage may need to approach the conversion to reduced and no-tillage carefully, because it may take time to learn the intricacies of a new system. Experimentation with various methods on smaller areas as opposed to going cold turkey with tillage may well be more financially stable.

This chapter covers field conversion methods, choices of tillage equipment, and reduced tillage, but these techniques and equipment are presented primarily as the means to achieve an eventual no-till system. As well, the chapter provides insights into how growers can use tillage in the correct place, at the best time, and with the most appropriate tools possible, and thus reduce tillage as much as possible. Likely there is no soil that would not benefit from an appropriate no-till system, as tillage is inherently unnatural and disruptive to soil biology and function.

No-till, in common agricultural terms, is often synonymous with herbicide usage, as the herbicide performs the primary function of elimination of existing vegetation. This chapter seeks to set the stage so growers need not use herbicides or tillage in order to achieve their vegetable production goals.

Clearing Woody Growth

Clearing existing perennial vegetation is an essential first step in preparing a field for annual vegetable production of any sort. This may be as simple as plowing in a sod, but in many regions, sod is mixed with scrub and tree regrowth. Some growers even face the daunting task of converting forest to cropland. When clearing trees and other woody vegetation, the general formula is chainsaw, remove the firewood, and haul off the tops and brush. Piled brush can be crushed with a tractor after a few years; this process greatly enhances the soil underneath. A heavy-duty mower serves well to reduce any residual vegetation after the clearing. If there is time to spare before the area must be brought into production, it's beneficial to then seed a cover crop of a perennial nature and mow it for a period of time; this allows stumps to begin to decay. A partially decayed stump pulls out with much less disturbance and more ease than a fresh-cut live stump.

Destumping can be very disturbing to the soil profile, so it is best to approach it as carefully as possible. Backhoes, excavators, tractors and chain, or ax and mattock can be used to remove stumps. Moldboard plowing of root-ridden, stumpy soil is very difficult. In fields with only small stumps, we have had success with not destumping upon clearing. Instead we used a combination of chisel plow and disk harrow to fit the field for vegetable crop seeding for a few years. Over time, the chisel plowing acted to pull out the roots and stumps. Of course there's the stones and boulders that needed to be removed, too . . .

When land is being converted to field for vegetable production, it is also the opportune time to address any major drainage projects. Before beginning the conversion process, it's ideal to observe the general water condition of a future growing area over a period of time—the longer the better, because groundwater characteristics can change dramatically over the course of a year. Drainage characteristics of a soil do improve in tandem with the general soil improvement, such as soil aggregation that the grower will facilitate. With careful observations, though, it often becomes obvious that a soil will need additional drainage efforts—for instance, soils in areas with seasonal high water tables. Drainage can be in the form of surface ditches or swales around the field area or subsurface drainage via stone-filled French drains or perforated pipe buried with stone. All drainage channels must discharge outside the field area, of course. When a backhoe or excavator is in the field digging drainage ditches, it is a prime time for a little careful destumping and boulder removal as well, if needed.

Tillage Tools and Techniques

Tillage tools range from plows and harrows down to hand shovels. Some are meant to work the soil very deeply, others to work only the surface. Tools that work the soil below a depth of about 10 inches (25 cm) are often called *subsoiling tools*. Tools for working the soil from a depth of a few inches to about 10 inches are *primary tillage tools*. Tools that penetrate the surface only a few inches are the *secondary tillage tools*. A traditional tillage routine is subsoiling and primary tillage in the fall if necessary, followed by spring secondary tillage. This allows the soil to recover its structure and biology to some degree through the winter months, with only surface disturbance occurring during

FIGURE 3.1. Hand tillage tools, *left to right*. Two-handled U-bar and the much stronger all-steel deep spading fork for depths up to 16 inches (40 cm). Spading shovel, spading fork, and two bed-preparation rakes (the backside of the smaller rake is particularly useful for leveling). The two tined hoes, *far right*, are used for breaking up surface compaction.

spring seedbed preparation. Regardless of time of year, tillage is best done when soils are relatively dry; this lessens compaction and soil structure deterioration in comparison with working wet soil. It is often said that when tilling, seeing a little dust flying off the soil surface is a desirable sign. This is especially the case with subsoiling and primary tillage and thus an additional reason to take advantage of any fall dry periods for primary tillage. Spring may not offer any such conditions in a timely manner.

PRIMARY TILLAGE TOOLS

The tools growers use to achieve primary tillage are the moldboard plow, the chisel plow, the offset disk, the rototiller, and the shovel. Primary tillage can achieve the incorporation of residues and fertilizer, the elimination of existing weed growth (particularly perennials), and influence the air:water ratio in the soil.

The shovel is the traditional gardener's tool for primary tillage. It is very gentle in action, allowing the user to invert sod carefully and control the depth of disturbance. The spirit of nature appreciates this level of human attention and effort. The shovel of choice for this job is the D-handle spading shovel, which has a flat, not pointed, bottom, giving an even depth of working. Working a large area, say an acre or more, with a shovel is generally not practical, but it should not be overlooked for smaller areas. Maybe in the future the shovel will make a comeback.

The moldboard plow is the tool of choice for quickly and thoroughly converting a thick perennial sod into loose soil for vegetable production. The plow's action inverts the sod, and a plowing depth of 6 to 10 inches (15–25 cm) keeps the inverted vegetation in the aerobic zone of decay while also reducing potential regrowth. Other tasks, such as terminating a cover crop, may warrant a shallower "skim" plowing of only a few inches deep. Though the moldboard plow is very effective for these tasks, it is also one of the most potentially damaging tillage tools. The inversion of soil creates a "wrong-side-up" condition where soil biology is also inverted, damaged, and exposed to drying. The moldboard plow is also famous for smearing the underlying soil below the tillage layer and creating a compacted plow pan. As well, the inverted vegetation can cause inconsistency in the soil structure, limiting moisture wicking from lower soil depths, and create toxic anaerobic decay gases if decomposition does not proceed smoothly.

The offset disk harrow is somewhat similar in impact to a moldboard plow. It is generally utilized for primary tillage once sod has already been converted by the moldboard plow. Like most disk harrows it is useful where there is an abundance of rock, boulder, or stone in the field. The offset disk harrow offers some inversion of materials and a thorough working of the surface of the soil.

FIGURE 3.2. Our tractor-mounted tillage tools include a moldboard plow, a tined rod weeder, and a water-filled drum roller. We use these when converting new growing areas to no-till.

A chisel plow is most frequently utilized by growers for subsoiling, but it is also utilized for primary tillage. It is capable of loosening soils without creating a plow pan. It introduces much air into the soil and provides a mixing of residues into the soil profile without full inversion. Like the offset disk it is commonly utilized for primary tillage in fields that have crops or cover crops already in place and not for the initial breaking of sod. It does not, however, provide a thorough working of the soil surface.

Vegetable growers commonly use rototillers for primary tillage. The rototiller is capable of thoroughly working the surface of the soil while incorporating residues. It is sometimes used to break sod as well. The rototiller is very effective at eliminating perennial weeds, especially with repeated passes separated by a week or so. This allows the perennials' exposed roots to dry out in the sun. Rototilling also incorporates much air into the soil. However, along with all its benefits, versatility, and ease of use, the rototiller causes the most damage of all the tillage tools. Its ability to destroy soil structure is extraordinary. As well, the soil underlying the rototilled layer is compacted by the smashing action of the spinning tines. The structureless surface layer releases its fine particles, which settle onto the compacted subsurface pan under rain or irrigation. The direct

FIGURE 3.3. A chisel plow can cut deep but produce minimal soil disturbance, even in a high-residue environment.

damage to soil life is also very significant. There is nothing gentle about this tsunami of destruction. Rototilled soils are almost always excessively aerated after the tillage event, yet they then collapse due to the destructuring, resulting in limited aeration. Rototilling has a dramatic impact on soil air, soil water, and soil temperature conditions.

SECONDARY TILLAGE TOOLS

Growers also use rototillers for secondary tillage or seedbed preparation, and it's possible to minimize the damage in these applications. For soils with minimal residues, or plowed or harrowed soils, rototilling at a depth of an inch or two can work to prepare the surface for the use of seeding equipment. Such tilling is often combined with use of a roller to smooth and repack the loosened surface. This can be very effective at eliminating an existing growing crop or cover crop and incorporating some of the residues.

The standard disk harrow is also a common secondary tillage tool. It is preferable to other tools following moldboard plowing due to its ability to quickly smooth a rough soil surface. It is capable of quickly working the full soil surface in a gentler action than the rototiller. However, depending on the type of surface residue present, a disk harrow may not be as thorough at eliminating weeds and cover crop growth as a rototiller. Use of a disk harrow combines well with use of other secondary tillage implements. The disk accomplishes a general chopping, flipping, and incorporation of residues, which then allows for a less aggressive follow-up with a lighter tool such as a rototiller, field cultivator, tined harrow, or even handheld rake. The disk harrow is also sometimes used to knock back vegetation or begin to incorporate residues before primary tillage with a moldboard plow.

Field cultivators generally have either points or sweeps mounted upon them (points and sweeps can be mounted on a tractor tool bar, too). Sweeps work well if a very thorough working of the surface is required (to eliminate young weeds, for example). Points are commonly used if the objective is simply to further prepare a seedbed that has gone through primary tillage or disk harrowing. The points of a field cultivator can be particularly useful for pulling out rhizomes of perennial weeds such as quack grass.

Tined harrows are the least aggressive tillage tools and generally are utilized for finishing a seedbed. They work the soil surface sufficiently to eliminate weeds and further smooth the surface. Before the proliferation of modern tillage equipment, the standard tillage routine was moldboard plow, disk harrow, and then the tined harrow for seedbed preparation. This can still be a very effective combination under certain circumstances. Tined harrows include the spike-tooth harrow, the rod weeder, and new types of harrows with thin tines. They are very gentle on the soil.

Hand tools for secondary tillage include tined or cultivating hoes and various rakes. Tined hoes can, surprisingly, be used to work large areas and may be suitable for refitting beds where only secondary tillage is required, probably not for acres of use but where a few beds need working. Being directly powered by humans, they are gentle in action and can be very thorough because of the proximity of the human. This can also be true of raking. People working with wide bed-preparation rakes can relatively quickly perform the final finishing of seedbeds. This would take the place of using tined harrows or rototilling under certain circumstances. Again, the raking sould be very gentle on the soil and is well worth considering where areas being worked are less than an acre or so.

Another tool invariably linked to tillage is the roller or packer. These tools are utilized to recompress the soil surface. They are often incorporated into bed-shaping equipment but have value beyond this. After tillage, the soil will benefit from being smoothed back to a level surface because it will help prevent damage to the soil life and limit erosion by water and wind. This is in addition to its use in reducing overaeration of a tilled soil profile. Rollers and packers range in form from simple wooden plankers to corrugated iron rollers or water-filled barrel-style rollers. Rollers are very useful at the final fitting of a bed for seeding and so are often the last tool employed in the tillage process. Rollers are also used again after seeding to ensure better seed-to-soil contact of broadcast crops or cover crops.

SUBSOILING TOOLS

Many tillage events cause compaction of the soil below the tillage layer. This is due not only to the action of the tillage tools themselves but also to the weight of the tractors used to pull the implements through the field. Unless tractors follow only permanent wheel tracks, they inevitably cross over the soil area upon which crops will grow. This tractor traffic often compacts the soil to a depth below that at which the tillage tool passes. Compacted layers or pans in soils are also very common due to various past human activities as well as natural conditions. Growers can dig into their soils and closely examine them for this issue, or use a penetrometer or similar tool for a quicker though less thorough assessment.

The grower can break up a plow pan through the use of additional tillage tools. The most common tools growers utilize for this effort are a chisel plow or a subsoiler. The shanks of these plows often reach 16 inches (40 cm) or more into the soil profile, ripping open narrow slots, which allows moisture to then readily move down through the soil profile. Eventually, once the soil restabilizes, the slots allow moisture to wick upward through the plow pan as well. The slots are often ripped about 3 feet (1 m) apart. Subsoiling does cause a significant destructuring of the soil in the slot, which takes time to reestablish. Because of this, it is preferable to subsoil in the fall and allow the soils to reconsolidate over winter before cropping with vegetables the following spring. Chisel plows are very capable of pulling up stones and even boulders, and growers need to be prepared to deal with such unearthed "gifts."

When it came time to acquire a chisel plow at Tobacco Road Farm, we simply purchased two chisel plow shanks to mount to our Farmall Super C tractor's existing 1½-inch (3.8 cm) toolbar. This was a much more affordable option than a separate chisel plow unit mounted to its own frame, and also avoided having another large piece of tillage equipment sitting around. The Super C is capable of pulling two chisels at a depth of about 16 inches in our relatively light soil. Wide beds receive two slots about 34 inches (85 cm) apart, the narrow beds receive one slot. When we were still practicing tillage, we subsoiled with a chisel plow every three years or so. The slot was placed in a different position in the bed each time chisel plowing was repeated. Now we chisel plow only when initially bringing a field into no-till production.

Another tool that is useful for the breaking of plow pans is the U-bar or subsoil digging fork. These are human-operated tools with long tines. The operator steps on the bar to drive the tines into the soil and then pulls back on the handles to move the tines through the earth. Their action can certainly be gentler on the soil than a subsoiler or chisel plow, but the human effort is substantial in comparison. The extent to which the handles are pulled back determines the level of soil mixing and aeration. We have used these tools to subsoil up to about an acre over the course of a season, with quite some effort. We were happy to replace them with a chisel plow, and even happier when the soils no longer required subsoiling.

A similar method for home gardeners is double digging or trenching. In this technique a spade's depth of soil is removed from a trench on one side of the garden and piled to the side that is not to be worked. The trench bottom is then loosened with a spading fork. The gardener then moves over and makes a trench alongside the first, essentially tossing topsoil over into the first trench. The second trench is then loosened with the digging fork. This process is repeated until the area is completely trenched and the soils are loosened. The initial removed soil is then brought over to fill the last trench. This is generally necessary only once when initially starting a garden or addressing a serious pan issue. It may be a useful method to quickly address garden soil structure difficulties.

If more time is available before a field must be planted with vegetable crops, cover cropping with deep-rooted plants can relieve plow pan compaction. In this case, plants with strong taproots are selected for seeding such as daikon radish, sweet clover, or burdock. Daikon radish is particularly fast growing, with late-summer sowings able to put on strong root growth before winter's arrival. The decaying radish offers a concentration of microbial activity, which helps to maintain an open channel into the subsoil. Sweet clovers take a little longer to establish but form very strong, effective taproots. As legumes, they assist the fixation of atmospheric nitrogen into organic forms in the soil. They generally flower in their second year of growth, after which the roots die back. Burdock is similar in growth pattern to sweet clover but forms very large, strong taproots that can effectively break up densely hardened soil pans. Many other cover crops also assist in the reduction of compaction. A combination of increasing the fertility of the soil through fertilization, eliminating or reducing tillage, and covering the growing area with a large mass of strong-rooted cover crops goes a long way toward a beneficial structuring of the soil.

Transitioning to Reduced Tillage

Often the first step toward reduced tillage is to set up a system that does not require primary tillage tools. Instead, the goal is to use secondary tillage tools for bed preparation with occasional subsoiling tools to break up compaction if necessary. This is particularly aided by the creation of more or less permanent beds and wheel tracks, so that heavy traffic and compaction do not impact the growing beds. Once a growing area has been converted to a vegetable production area, it is primarily machinery traffic, along with tillage itself, that results in conditions where primary tillage will be necessary. Permanent bedding is generally created by having a consistent width of bed and wheel track to match the appropriate tractors that will straddle this bed surface. The wheel tracks between beds are often depressed by the weight of the machines, with the bed potentially being raised through hilling with the soil from the wheel track as well. These depressed wheel tracks make it easier to see where the wheel track is as the crops grow. If a wheel track becomes obscured, it can be reestablished by noting the location of the neighboring wheel tracks. Though the soil may become compacted under them, permanent wheel tracks do not require any tillage since they will not be part of the growing area. Secondary tillage tools can be selected to match the appropriate bed width. Secondary tillage might obscure wheel tracks to some degree, so bed reshaping or hilling may also be necessary depending on the tillage tool.

Permanent bedding systems go a long way toward reducing the need to use the more damaging primary tillage tools, as well as reducing the area worked by tillage tools. To further reduce the need for secondary tillage tools, effective weed control and techniques such as flail mowing of crop residues can be of assistance. The reduction in material size by mowing before incorporation can allow use of the lighter tillage tools like harrows and cultivators for seedbed preparation, thus reducing the need for the soil damage of the rototiller. On Tobacco Road Farm our conversion from a tillage-intensive system began with the establishment of permanent bedding and the abandonment of broadacre primary tillage. Beds were then individually rototilled (shallowly) and chisel plowed when necessary. Bed shaping immediately followed using tilling disks to move soil up out of the wheel tracks and a roller to finish the surface. With further refinement we were able to move away from the rototiller to surface preparation with disk harrow and field cultivator, and eventually moving into no-till. This occurred over the course of several years with plenty of trials and experimentation to lead the way.

BED LAYOUT CONSIDERATIONS

Many factors related to field layout can impact efficiency and thus profit. The layout of the bedding system is very useful for helping to balance field conditions, for example. The direction in which beds are laid out can influence the impact of the elemental forces upon crop growth. The raising of beds or hilling of crops likewise influences these forces. Again these are primary concerns for achieving the balance of forces for vital crop growth. Along with appropriate planning, crop selection, and rotation, thoughtful field layout helps growers achieve the goal of a bountiful crop.

FIGURE 3.4. These beds are laid out against the slope to limit erosion, but still allow excess water to run off.

When considering field layout it is important to balance the need for efficiency with the environmental needs of the crops. For instance it may be most efficient to set up growing areas close to the packing and processing area for speed of harvest. Yet the soil, light, water, or air conditions for a specific crop may be more favorable in a location farther from the processing area. Another example of this is whether to embark on production in two or more physically isolated fields. On the one hand physical distance between fields can reduce certain efficiencies; on the other hand fields in physical isolation lend themselves very well to practicing crop rotation to avoid insect damage and disease.

When considering how to lay out beds in a field, the primary concerns are soil, water movement, airflow, and sunlight. In general as long as ample sunlight is available, vegetable production beds are best laid out perpendicular to a slope to avoid soil erosion by rain. There is also the additional concern of positioning beds to encourage sufficient airflow through a crop, or to limit airflow depending upon conditions. Designing bed placement with these considerations as paramount may not result in the most efficient field layout in terms of labor and the movement of humans from place to place, yet when the whole picture is considered, it yields the best overall productivity.

Vegetable crops are often grown in bedding systems in which relatively wide growing beds are not utilized as a walkway or driven on by mechanical equipment. In row cropping systems, crops are grown in single rows at a standard distance, and foot and wheel traffic is acceptable between any and all rows. Potatoes and sweet corn are common examples of crops often grown in rows.

SETTING BED WIDTHS AND LENGTHS

Bed width is determined largely by the equipment that a grower will use, particularly tractors. In one field at Tobacco Road Farm, we laid out 36-inch (1 m) wide beds separated by 8-inch (20 cm) wide wheel tracks, which works well for using low tunnels. We find that relatively narrow low tunnels are better able to withstand snow load as well as wind difficulties, and they work well with our International Cub tractors. The narrow 8-inch wheel track is just enough width to walk down, with crops often growing to cover it over for maximum field solar collection.

In our other fields, the beds are 58 inches (1.5 m) wide with 10-inch (25 cm) wide wheel tracks, 68 inches (175 cm) total from bed center to bed center. We do not use low tunnels in these fields. This wide layout allows us to convert these beds into row cropping areas for the potato crop, with two 34-inch (85 cm) wide rows taking the place of the bed and wheel track. Our choice of machines to work these beds or rows are Farmall Super C tractors. These tractors are able to pull bigger equipment such as manure spreaders, potato diggers, and chisel plows through the field. One of the Super C tractors has a narrow single-tire front end and extra-wide rear axles. The narrow front end gives an excellent turning radius. It also fits over two of the 36-inch-wide beds, with the rear tires placed 88 inches apart center to center, and the front tire running down the middle wheel track, giving us additional traction when needed in the low-tunnel field as well. It is not usable on 58-inch-wide beds, however, it can be set up for use in the 34-inch row crops.

We maintain two sets of bed-shaping equipment and light tillage tools for these bedding systems. There are 36-inch and 58-inch rollers, disk harrows, and tined harrows, as well as the appropriate belly-mounted cultivation equipment. There is a mounted dump cart on a Cub tractor for compost delivery onto the 36-inch beds. The manure spreader and dump truck fit the 58-inch bedding pattern.

For row or bed length, it is very useful to set a standard length. This simplifies crop area calculation, setup of irrigation equipment, and management of row covers and solarization covers, which are key components of our no-till system (described in detail in chapter 4). If all beds are the same length, row covers and solarization covers can all be cut to the same length and are therefore interchangeable. For our low-tunnel field, bed lengths are multiples of 40 feet (12 m).

We chose 40 feet because that is about the maximum distance that two people can effectively lift and place a solid plastic row cover over low tunnel support hoops without having to walk along the bed length. In our other fields, beds are approximately 150 feet (45 m) long. We cut the plastic covers to about 50 feet (15 m) in length; that way they can serve not only as low tunnel covers but also as solarization sheets for any of the beds in any fields (three covers per bed for the 150-foot beds).

The bedding system in these fields is permanent; in other words, the wheel tracks remain in place from year to year. Even when we convert 58-inch-wide beds to a potato row cropping scheme for a season, the tractor tires still run along the same wheel tracks, with an added wheel track down the middle of the wide bed. Every few years the growing beds may need to be reestablished by using the bed-shaping equipment so that we can accurately distinguish where the wheel tracks are. The bed-shaping equipment is simply large disks mounted on the belly cultivation bars, with an appropriate-sized roller following on the rear hitch. This equipment creates a slightly raised bed, which eventually flattens out again over the following few cropping cycles. Raised beds are not essential in no-till systems; the decisions as to whether to create raised beds is dependent upon conditions. If conditions are watery and cold, then raised beds may be of benefit; if conditions are airy, hot, and dry, they can be detrimental.

Direct Conversion of Sod to No-Till

If you are fortunate enough to have a field of sod, such as a hayfield, with no woody plants to contend with, it is possible to convert that sod to no-till readiness in a relatively short period of time. This is accomplished by mowing the vegetation, turning the sod, fully eliminating perennial weeds, and shaping beds and cover-cropping in fall, followed by solarizing and no-till planting the following spring. This could also be accomplished over the course of the spring. Here is a description of a tillage system that we have utilized to accomplish this.

Starting with hay sod, we mow down the vegetation in the fall, preferably using a flail mower that chops up the residue; a disk harrow could also serve this purpose. If the field needs extensive soil amendment with limestone or other mineral application, materials can be applied to the soil surface for incorporation by tillage at this point. After the chopped vegetation has dried for a few days, we select dry weather to begin tillage. The first tillage is a careful moldboard plowing to fully invert the sod, followed on the same day with a disk harrow to smooth the plowed surface. The moldboard plow and disk harrowing are broadfield tillage, meaning that the tractor tires are not restricted to a certain area or pattern on the field. Once beds are formed tillage will be confined to the bed width.

In order to shape beds, the belly-mount tilling disks on the tractor toolbars with the rear-mounted roller are utilized by starting at one side of the tilled field and then forming adjacent beds as determined by the placement of the previous wheeltrack. This shaping and rolling goes a long way toward preserving soil quality, so if possible plowing, harrowing, and bed shaping all happen on the same day.

The next step is chisel plowing to relieve the compacted layer that is inevitably in place on tractor-harvested hay land as well as any compaction from the primary tillage event. As mentioned above, our beds are 58 inches or 36 inches wide. We use two shanks on the wider beds and one on the narrower beds. We remove any dislodged stones and boulders using the same tractors that fit the bed shape, for the 58-inch beds a Super C. For this the Super C is mounted with a platform on the rear lift to pile the stones on. If a boulder is too large to push onto the platform, it is chained and dragged out of the field, followed by a bit of raking to resmooth the bed. If the chisel plow pulls up many rhizomes such as quack grass, we wait a few days to allow the rhizomes to dry out. (In cases where living rhizomes persist even after a drying period, we run the Super C with belly-mounted cultivator shovels over the bed every week or so to uproot them to a satisfactory degree.)

We might finish up with a pass of the rod weeder (tined harrow), then broadcast cover crop seed and roll for better seed-to-soil contact.

The following spring we mow and solarize the cover crop (solarizing is described in chapter 4). As long as perennial weeds are basically fully eliminated, we would then move right into no-till. If perennial weeds are still viable, we rototill to an appropriate depth, or make additional use of the field cultivator, with appropriate breaks to allow the weeds to dry out, until the perennials are sufficiently destroyed. At this point the beds are ready to seed or plant, and the no-till adventure begins.

CHAPTER 4

No-Till Techniques

A pesticide-free, no-till approach to vegetable growing brings the structure and biological conditions of the soil to a high level of function, which, along with appropriate soil fertility practices, reliably produces vital, abundant crops. With such soils in place, weather and environmental challenges can be overcome, crop production is efficient, profitability is likely to be maintained, and grower and customer satisfaction is high.

Fertile soils left undisturbed by tillage or pesticides do not develop surface crusts or plow pans. The soil air, water, and temperature conditions are well maintained. The soil biology thrives under such conditions, with all the benefits described in previous chapters: improved soil aggregation, increased organic matter, rapid decomposition of residues and humic compound development, and balanced nutrient release to crops. The improvements to structure and biology are synergistic: Better structure means more soil life, and vice versa. These improvements lead to efficiencies in grower efforts. The short list of improvements that arise from no-till management includes greater ease of planting, less need for irrigation, far fewer weeds to be controlled, crops that show better disease and pest resistance and increased cold tolerance, speedier harvests, and higher-quality produce.

As discussed in the previous chapter, many steps are involved in bringing a field into the conditions where no-till vegetable culture can be initiated and successfully perpetuated. Once those preconditions have been met, it is time to begin moving into full non-soil-disturbance techniques (no-till). Developing an effective system to end the growth of a previous crop without tillage has traditionally been the most challenging area of no-till production. In many no-till agricultural systems, herbicides are generally used to kill the previous crop, or cover crop, or weeds. In the no-till system we have developed at Tobacco Road Farm, however, we use no herbicides. Instead the three main steps in our system to prepare beds for seeding or planting are mowing whatever plant material is

59

growing in the bed, eliminating regrowth of the living vegetation that remains after mowing, and application of compost and mulch material.

Mowing effectively transforms the existing top growth of plants into a mulch material. The roots of the freshly mown plants are weak but are often capable of regrowth, so it is best to immediately address that possibility, because regrowth would interfere with planting the bed and development of a new crop. The five main techniques to eliminate regrowth are solarization, mulching, exposure to cold temperatures, hoeing, and mowing during the flowering period. The appropriate technique for elimination is determined by the time of year or environmental conditions. In addition, applying compost at this time also assists with the smothering of vegetation, and it helps in preparing the bed surface for successful seeding or transplanting.

Mowing

Mowing is a highly effective way to reduce existing vegetative cover, whether a standing cover crop, leftover vegetable crop growth, or just weeds. Mowing reduces the height of existing vegetation and lays it on the ground surface, where it will begin decomposition.

The selection of the mowing equipment depends on the objective and conditions. The mowing machine we use most often at Tobacco Road Farm is a two-wheeled BCS Model 850 power unit with a front-mounted flail mower (see figure 4.1). The 850 is a large, older BCS with a 16-horsepower engine. The BCS is light enough to minimize soil compaction, and its maneuverability is superior to that of our tractor-mounted mowing machines. The flail mower has vertical spinning flails that are very effective for coarsely chopping crop residues, and it is capable of chopping thick growth without clogging. It is our mowing implement of choice for most applications, but the BCS can also be mounted with a sickle bar mower or a rotary mower. The sickle bar mower has blades that move in a horizontal back-and-forth pattern. The rotary mower is essentially a heavy-duty lawn mower; its blades spin horizontally.

The implements can be mounted with a quick disconnect which makes implement changeover less strenuous, and it's easy to remove the blades for sharpening on a grinding wheel. Keeping the blades sharp greatly improves cutting efficiency. Removal of stones from the bed's surface before mowing goes a long way toward keeping blades sharp as well. We distribute 5-gallon (20 L) buckets with holes drilled in the bottom around the field edges for collecting any loose stones. Generally stone removal becomes much less of an effort once tillage is discontinued, but some may still end up on bed surfaces by way of top-dressed compost or mulches. The collected stones are useful for road improvements or in the base layer on which compost piles are assembled.

FIGURE 4.1. We mow using this BCS two-wheeled tractor and various attachments: a flail mower (mounted on the BCS), the undercutting sickle bar mower (*foreground*), a rotary mower (*left*), and a rototiller (*back*). We also make good use of modern lightweight handheld scythes like the one leaning against the BCS.

The flail and rotary mowers are 30 inches (75 cm) wide, which allows us to mow 36-inch (1 m) beds with 8-inch (20 cm) wheel tracks on either side (52 inches / 1.3 m total) in two offset passes, or three offset passes on the 58-inch (1.5 m) beds with 10-inch (25 cm) wheel tracks on either side (78 inches / 2 m total). Mowing the wheel track area is only necessary if weeds have grown on them. Otherwise only the bed surface needs to be mown. The flail mower is particularly effective at reducing vegetation in a single pass. Occasionally, if the vegetation is very tall or thick, we find it effective to press down on the handlebars to lift the front of the mower. This prevents dragging or overworking the implement, but it can be laborious to sustain this technique over a large area. This thick vegetation then requires another pass at the standard mower height.

To best assure the death of the stubble following mowing, we cut quite low, about 1 inch (2.5 cm) above the soil surface. This particularly helps ensure faster, effective results during the solarization step of bed preparation. Our slightly raised beds have a gentle sloping edge, and the flexibility of the relatively small BCS mower effectively follows the bed and field contours, resulting in that "crew cut" we're aiming for. Another benefit of these smaller mowing machines is the ability to avoid mowing leaping toads, fleeing snakes, and other beneficial organisms. Not only is the mover operator closer to the ground than when riding on a tractor, but with the BCS it is also easier to more quickly make changes in speed and orientation of the mowing implement. The sickle bar mower is the gentlest of the BCS mowing implements in terms of danger to beneficials.

The flail mower lays down the chopped residue in place on top of the bed. The rotary mower, in contrast, blows a substantial portion of the mown materials to the sides of the machine. The discharge vents sometimes clog with residues. Grandmother did always say that "patience is a virtue," and we quickly clean out the vents, and perhaps aim to mow a little less material on the next pass. If there is a crop growing alongside an area that needs to be mown, an option is to seal one or both of the side discharge vents. This prevents the mown materials from landing on leaf surfaces of the adjacent crop and blocking photosynthesis or becoming annoying debris in a crop that is close to harvest. In some cases it is useful to blow residues under a neighboring crop; one example is corn, where the residues will contribute to fertility and weed control. Note that when both discharge vents are sealed, the machine may struggle with thick vegetation, requiring a slower pace of work.

FIGURE 4.2. We use the BCS flail mower to cut down dense cover crops before solarizing.

The rotary mower has a bagger that fits atop the implement. This is useful to collect seed heads if the residues are weedy. To prevent the weed seeds from becoming future weeds, the bagged chopped materials are removed from the field and dumped into weed piles on the field edges. These materials, once decomposed, can be spread underneath fruit trees and other perennials as a fertility material. Of course, once a no-till system is functioning effectively, there will seldom be weeds in seed in the growing beds. If there are just a few weed seed heads in a bed, they can be removed by hand before mowing using a machete or knife, or possibly a hoe for low-growing weeds such as chickweed. This is also a time to remove any perennial weeds, as they are generally resistant to quick solarization efforts. If the soil has come into a state of much aggregation, many perennial weeds can be simply pulled out by hand,

FIGURE 4.3. The residue of a winter rye / crimson clover cover crop covers the soil surface of a recently solarized area in the *foreground*. In the *middle* area covered with plastic, solarizing is ongoing. In the *background* more rye and clover await mowing.

roots and all. Other conditions, or particularly difficult roots such as Canada thistle, will require trowels or trenching shovels. The effort of weed seed head removal pays off tremendously in the overall progression from a weedy to a weed-free soil surface.

The mown residues are left on the soil surface to decompose, which feeds the soil biology and provides a mulchlike layer that helps diminish weed seed germination. If the residues constitute a layer thick enough to cause subsequent top-dressed compost to excessively dry or to impede seeding, some of the residue may need to be removed. How much residue a given soil can handle depends on the nature of the materials and how hungry the soil biology is. With a highly functioning biological system, residues are consumed shockingly quickly, especially when some compost is also applied to the surface after mowing. The nutritive content of the residues also impacts not only decomposition rates but also future fertilizer decisions. For instance, a bed covered with residue from a lush cover crop of nitrogen-rich legumes will require different fertilization for balance than one covered by a decay-resistant, carbon-rich residue of winter rye mown at flowering.

The selection of the appropriate mowing implement is often related to the desired speed of residue decay, which will depend on how long soil surfaces need coverage and how quickly nutrient release from the decomposing material is sought. Chopped residue from the BCS flail mower usually consists of pieces

FIGURE 4.4. A thick mulch remains after solarization of a cover crop primarily composed of winter rye. It can be raked aside for compost application and seeding and then placed back on top, or transplants can be slipped into place right through the mulch. Most crops would require significant side-dressing later on if planted without added compost.

only a couple of inches in length (see figure 4.4), so it decays fairly fast. The rotary mower finely chops up vegetation, making it available for quick decay. Front-mounted on the BCS, the sickle bar mower works well for undercutting tall vegetation, such as some dense cover crop mixtures, in a single pass. This mower leaves the cut plant tops intact, rather than chopping them up. The long-stemmed residue is slower to decompose, which may be of benefit under certain conditions such as delaying crop residue nutrient release or providing longer-term soil cover.

The tractor-mounted versions of these mowing machines are commonly used for these tasks on other farms and may be appropriate for large areas. For smaller areas even an inexpensive old lawn mower may serve for mowing crop residues. An even gentler approach to mowing is the use of hand tools. With 3 acres (1.2 ha) of vegetables in cultivation, we mostly utilize the BCS mowing machine, but occasionally an area is so small or isolated that we cut it with a machete, scythe, or sickle. All of the tools need to be kept quite sharp to be effective, and attention must be paid to mowing low enough. A few crops—tomatoes, peppers, trellised pole beans and peas—cannot be mowed with machines for reasons such as the presence of twine used for trellising. In this case we cut down the trellising first and then cut the plants at ground level with machetes. A special hand tool that looks like a cross between a machete and a hatchet (purchased from Fiskars) is our tool of choice for dislodging stems and some of the upper roots in hard-to-decompose stubble such as eggplant stems.

Solarization

Solarization is the use of plastic sheeting on the soil surface to capture solar energy and eliminate regrowth of mown plants. This technique is highly effective if the temperature under the plastic rises sufficiently. Clear plastic sheeting is our material of choice because it heats faster and hotter than black plastics under sunny conditions. The ambient temperature needs to be about 75°F (25°C); slightly lower temperatures may be sufficient depending on conditions. Solarization is reliably effective at these temperatures in Zone 6 Connecticut from May through September. We have had some success in April and October warm periods. The day length and declination of the sun in spring support a bit more temperature gain under the solarizing cover than in fall.

The type of residue being solarized also impacts speed and effectiveness. For example, mown lettuce residue will die more easily than a heavy winter rye cover crop. Large amounts of mown material lying on the bed surface can also increase the time required for effective solarization. The height of the stubble is significant as well. If temperatures are hovering around 75°F on a May day, and the intention is to solarize a heavy cover crop of winter rye, the mower may need to be set quite low, an inch or even less above the soil surface. If you are solarizing a vegetable crop in 95°F (35°C) July heat and sun, roughly knocking down the crop enough to spread a cover over it with sides secured to the ground will probably suffice. Under the solarization cover on a sunny day, the air temperature increases to about 50°F (22°C) hotter than outside air temperature; so on an 80°F (27°C) day, the temperature at the soil surface under the cover could be 130°F (54°C). Soil temperature increases, too, but not as much. The gain at a depth of 1 inch (2.5 cm) is only about 10°F (6°C) degrees above ambient air temperature. In order to avoid damage to soil biology, the covers are left in place for as short a time as possible. One day of solarization is usually sufficient in the summer, but it can be a matter of only a few hours under hot conditions or up to several days under cooler or more difficult conditions such as cloudy weather or thick mown residues. We monitor the covers for effectiveness and will gladly pull them off a little early and hoe off the few straggling survivors rather than leave the covers in place for more time and risk harming soil biology. The roots of established perennials, however, are relatively resistant to quick solarization. This is the primary reason why farm fields and garden beds need careful, thorough preparation to remove perennial weeds before moving into no-till: Clear plastic solarization is our primary means of killing off existing vegetation before seeding or planting, and it is *not* effective against perennial plants. This also speaks to the importance of carefully selecting the materials used for mulching beds and building compost piles in order to avoid bringing new perennial weeds into your production areas. (Mulching and our compost-making process are discussed in detail in chapters 6 and 10.) For perennial weeds that do enter into the no-till production area, it is often easier to notice them among the crop residues before you mow, though some can be seen clearly even after mowing. Either way it is well worth the time it takes to remove those weeds, uprooting them by hand, trowel, or shovel, before you put solarization covers in place.

MANAGING THE COVERS

We often reuse covers from high or low tunnels for solarizing beds. It's important that the covers have a minimal number of holes, though. The high tunnel covers are 6 mil and measure 32 × 50 feet (10 × 15 m). The low tunnel covers are 2 mil and measure 8 × 50 (2.4 × 15 m). Another option is 100-foot (30 m) lengths of purchased 4-mil construction-grade plastic of various widths; these covers are often cut into 50-foot lengths. The standard 50-foot length is useful for our bed length, which is either 40 feet (12 m) or 150 feet (46 m). One length or three lengths of covering is sufficient for any given bed. The high tunnel covers and construction-grade plastic are the better choice for solarizing larger areas because they are wider. They may need to be applied in groups. A 6-mil 32 × 50 plastic sheet is relatively heavy to drag about and is about the maximum size one person can readily handle. The 2-mil 8 × 50 low tunnel covers work well for smaller areas. These covers are much lighter and easier to work with and are covers of choice in our diversely planted year-round production field. The wider covers find more use on the fields where larger amounts of a single crop are seeded at once. In this case we often cover ¼ acre (0.1 ha) or more at a time.

Sandbags placed every 20 feet (6 m) or so along the edges hold the solarization covers in place. We use sandbags made of black 6-mil UV-treated plastic. If an area needs multiple covers, we overlap the edges by a couple of feet and then secure them with sandbags. It is important to limit the access of air under the cover—mowing low, securing the sides well with the appropriate number of sandbags, and limiting the number of holes in the cover all work in favor of effective solarization. If the covers are wide and windy conditions occur, a few sandbags may need to be tossed onto the middle of the covers. We park extra sandbags along the field edges in small piles when not in use. Sometimes when bed preparations are advancing quickly, we simply move sandbags from a solarized area directly to the next area to be covered.

Seeding or transplanting needs to immediately follow a solarization event, so it's important to limit the amount of field area being solarized at any one time to a manageable amount. At Tobacco Road Farm we may mow and solarize as much as ¼ acre at a time. Once the stubble in that area has died, we mow the next ¼ acre and shift the bags and covers. We compost, seed or plant, and mulch the first ¼ acre in the day or two that it takes for the next area to solarize. In this manner we work across the field and require only a ¼ acre of plastic sheeting to maintain a steady flow of work. Of course, with a larger crew and additional plastic and sandbags, more could be accomplished in the same period of time.

Solarization in this manner is shockingly effective at creating a completely browned-off surface layer with no annual vegetation surviving. After almost 10 years of observation and trials of short-duration solarization use on our farm, it does not appear that soil biology is significantly damaged by these quick exposures to high temperatures. The physical characteristics of the soil, such as aggregation and biological intensity, have continued to improve, and crop health is excellent. In this no-till system, the soil remains well structured and not overaerated. It also stays well hydrated and has a thoroughly mulched surface. All of these conditions would limit damage to the biology from increased temperature. A bare, tilled, dry soil may not fare as well under such a solarization event.

LONG-TERM SOLARIZING OF PERENNIALS

Our solarization technique differs from the traditional use of clear plastic to sterilize soils against pathogenic soil organisms. Sterilizing soil requires leaving the covers on for several weeks or more. Of course we are not interested in sterilizing our soil biology; rather we seek to enhance it so that pathogenic conditions are not dominant. There is also more risk that plastics (another potential soil contaminant) will leach into the soil with long-term coverage. Long-term coverage of soils with plastics may also inhibit the ability of the soil to effectively breathe.

Despite some of the negative attributes of long-term plastic usage, occasionally there may be conditions where its use is warranted. For longer-term solarization black plastics are utilized for their ability to block light, as well as to heat up under the sun. This ability to block sunlight makes black plastic rather than clear plastic an option that may help solve a persistent perennial weed difficulty, prepare beds by eliminating annuals during the cooler months, or control heat-resistant annual weeds such as purslane. The cover will need to remain in place for weeks to months depending on the plant species being covered. Here in Zone 6 Connecticut during the warmer period of late spring through early fall when temperatures can exceed 90°F (32°C), many of the perennials or tough annual succulents such as purslane may take an average of three to four weeks to die out—less time if temperatures are very high, or more during cooler periods. Very difficult perennial weeds like Canada thistle may take considerably longer.

Using black plastic can also be effective for killing sod or other perennial vegetation to start a vegetable-growing area, but it does not address potential pre-existing fertility, plow pan, or compaction issues. These are the primary reasons we utilize tillage to bring an area into no-till production. In addition, the soil biology is potentially damaged during long periods of covering, especially during hot weather, and the soil does not receive the benefit of root exudates because the covered vegetation is not photosynthesizing. As well, there is increased potential for plastic leaching into the soil, especially with high summer temperatures.

During the colder time of year, many hardy plants are particularly resistant to being starved of light. This characteristic has likely evolved so that the plants can survive long periods of being light-deprived under snow. It may take five to six months to fully kill out vegetation under covers that are spread over beds in the fall. Covering during this period can be very effective for controlling perennial weeds that put out growth in the early spring.

Our black plastic of choice is a 6-mil silage tarp. The material is black on one side and white on the other. We place them with the black side faceup, but it's possible that laying the tarps with the white side faceup might be less damaging to soil during the high heat of summer. The silage tarps we have worked with are well made and resistant to decaying in the sun. We have also used 4-mil construction-grade black plastic, but its resistance to decay is much more variable, and it is irritating to have to pick up large volumes of small shards of black plastic sheeting. Other growers have successfully utilized black landscape fabrics as well as the thin black plastic sheeting commonly employed as a seasonal cover for growing beds.

One important technique to help avoid soil damage when using black tarps long-term is to punch a few holes in them to allow water to penetrate and keep the soil hydrated. The distance between these holes can be quite variable depending upon soil conditions. Because of its superior structure, a no-till soil is much better prepared to remain hydrated through appropriate wicking. Therefore the holes can be much farther apart, maybe every 15 feet (5 m). A tilled, destructured soil may respond better with significantly closer spacing. When we purchase a new silage tarp, we use it for covering materials such as straw; as the tarps develop holes through repeated usage, we eventually shift them to use for long-term solarizing, and we simply add a few extra holes as needed.

After solarization is finished and we remove the tarps, if the residue is excessive, we move it into the wheel tracks, temporarily onto a neighboring bed, to the edge of the field, or to a composting area. The most common excessive residue is a heavy rye cover crop. This is a useful straw material for mulching

FIGURE 4.5. Black plastic covers spread in the fall to remain in place through the winter. When we later uncovered this area in the spring, we found that the cover had effectively eliminated a patch of Canada thistle.

around transplants or on top of broadcast seeded beds, so we usually stockpile it nearby the growing areas. To move the residues we often simply rake the bed surface with bed preparation rakes that have wide, adjustable heads. If the area to be raked is large, we may use rod weeders mounted on the fast hitch on a tractor to gather the material into piles for quicker removal via carts to the field edge. Usually it is easy enough at planting time to simply hand-rake material a bed or two at a time, transplant or seed into those beds, and put the material back into place as a mulch.

More Smothering Techniques

A similar approach to using black plastic is to eliminate light by spreading mulch material over the mown vegetation. This method is very friendly to the soil biology, because it provides a food source, involves few or no toxins, and presents no risk of damaging heat accumulation. However, a mulch layer alone may not stop tenacious perennials from growing. Mulch is most effective on top of mown vegetable crops of relatively weak regrowth potential. (Many vegetable crops are of such nature.) It also works well to smother newly germinating weeds in growing areas in late fall when it is too late to seed a vegetable or cover crop. For direct seeding, sufficient time needs to be allowed for the buried vegetation from a previous crop to die out; the mulch is then raked aside or removed from the bed being prepared for seeding. Transplants can be set through this mulch without raking. Heavy mulching is a relatively common approach for keeping unplanted beds covered in winter.

In order to stop regrowth of vigorous weeds such as quack grass, a layer of cardboard may be laid on the soil first, and the mulch spread over the cardboard. The cardboard is often effective at blocking growth of rhizomes and is quite popular with soil biology as a food source. Cardboard takes some time to decompose, so beds are not usually fit for reseeding for a period of months. An alternative is to poke holes through the cardboard and set transplants.

Cardboard can also be utilized to cover perennial vegetation in order to convert an area to vegetable growing, but as with black plastic, this technique may fail to address compaction and potential fertilization needs. It can be useful for starting a new garden of limited size.

Using hot mulch materials is another effective approach to eliminating regrowth. This approach is similar to building an on-site compost pile that effectively heats the soil surface, denies light, and provides a physical smothering of material, thus blocking the growth of even tenacious perennials. This material may be as simple as a 2-foot (60 cm) or deeper layer of fresh-cut hay or grass clippings. This nitrogen-rich material is capable of heating to 150°F (65°C) or more without the addition of any other material. Carbon-rich materials such as sawdust, leaves, straw, and wood chips can also be effective if a nitrogen source is added in appropriate proportion. This mulch can heat quite hot; time may be required to allow it to cool before transplanting can be commenced. These heavy mulches generally need to be raked aside before seeding a bed. We have successfully used this method between rows of winter squash vine with a 6-foot (2 m) row spacing in order to burn out a quack grass infestation. The heat and nutrients released from the fresh-cut hay mulch also boosted the squash yield.

Another approach to eliminate a previous crop is simply to let cold temperatures kill it off. Mow down frost-sensitive crops in the fall when freezing temperatures are approaching. This would include the late-season squash crops, cucumbers, tomatoes, peppers, eggplants, corn, callaloo, and more. The stemmy stubble left behind is simply composted over and the beds reseeded. Approaching cold weather ends any potential regrowth. This works best when such crops are growing in beds with effective weed control, as weeds may not be killed by the freezing temperatures. This approach is also useful with "half

FIGURE 4.6. Hot mulch! Sawdust, *bottom*, had nitrogen-rich fertilizer applied to create heat. This then burned out a patch of quack grass.

hardy" crops and cover crops that will eventually die out when exposed to low winter temperatures. We commonly utilize this approach with fall-seeded cover crops of oats, barley, and field peas. The death of these cover crops by cold temperatures provides a mulched bed surface that can be transplanted into as early as late winter. Before seeding crops, the residue may need to be raked aside, then reapplied as a mulch. It is also possible to expose half-hardy crops like lettuce growing under low tunnels by pulling the cover off to the side, which allows the cold winter temperatures to end the crop. Then the bed can be seeded with another crop and the cover reapplied. For these approaches, if any winter-hardy weeds are present, it's important to hoe them out or otherwise remove them before establishing the next crop.

Many vegetable crops that are grown in the winter will not readily succumb to death by cold temperatures. From October to April—our cold-weather season—we often hoe out vegetable crop residues, especially on our acre of low tunnels where residue of harvested crops often needs to be removed expeditiously to make way for the seeding of the next crop. This is usually not difficult work, because there are few to no weeds and winter vegetable residues are relatively weak. If the residue is very thick it may be mown before hoeing. We use 10-inch (25 cm) wide DeWit half moon hoes or, if the residue requires more aggression, the field grub hoe made by Rogue Hoe. Hoes are kept very sharp by the use of a grinding wheel, and the entire bed surface can be effectively undercut by careful hoeing technique. If the residue is excessive enough to get in the way of the next seeding, we rake it off and compost it, or set it aside in weed piles if it contains weed seed heads, such as chickweed or dead nettle.

Yet another option in eliminating a previous crop is to mow or roll and crimp at a specific period in its growth cycle. This is a somewhat common approach on broadacre field crops in no-till systems that do not rely on herbicides. Often a roller/crimper front-mounted on a tractor is used to crush the cover crop at its flowering stage. A no-till high-residue seeder specifically designed for this purpose is then pulled through the field to seed a crop such as corn, small grains, or legumes. We have experimented with this approach in our vegetable system with limited success. The first difficulty is waiting for the cover crop to be at just the right stage of maturity for effective elimination. This rolling/mowing timing is not versatile enough to coincide with appropriate seeding or planting timing for many vegetable crops. The second difficulty is that although mowing or rolling/crimping may effectively kill the cover crop, it is much less likely to eliminate any weeds that are growing amidst that cover crop. The third difficulty is the slugs—slugs often thrive in an environment with lots of decomposing residue. By contrast the mowing/solarizing method largely addresses all three of these critical areas and provides us with great flexibility in our fast-moving vegetable system. Occasionally we have mowed a weed-free crop or a cover crop in its flowering state and it did not regrow, allowing us to skip the solarization step. Occasionally crop harvest is so thorough that solarization is not required to prepare the bed for the next crop. This is most common after a root crop harvest (garlic, potato, carrot, beet, turnip, radish, and more) in a weed-free growing area. In this case when the harvest of the

FIGURE 4.7. This mown amaranth died off on its own with the onset of cold weather. There are no weeds growing, either, so no need to solarize this bed before planting the next crop.

FIGURE 4.8. After garlic harvest there are no weeds. We simply rake aside the excess mulch and seed the next crop.

crop is complete, the growing area may be raked if residue is excessive, followed by a quick hoeing to finish off any straggler weeds or crops, and then planted.

Other crop elimination techniques that we have experimented with include organic herbicide sprays and flaming. The organic herbicide sprays contained concentrated vinegar as their most active ingredient. During sunny, warm conditions the vinegar herbicide was effective at eliminating small green vegetation, but what we most often need to eliminate is older, tougher residues. As well, the significant cost and effort far exceeded those of mowing and solarizing.

Weed size is also an issue when flaming, which is more effective at dealing with small weeds. It is not particularly useful against larger residues. Also there is the issue of all the mulch on the field, which does readily light on fire. We have experience with that. Flaming proved useful on one occasion when a lot of chickweed came up through the mulch among young lettuce seedlings. We thoroughly wetted the mulched bed, then quickly flamed off the lettuce and the chickweed seedlings. This then provided us a weed-free bed surface. The wet mulch protected the soil from the ravages of the flame, and we were able to broadcast new lettuce seed right into the mulch without additional chickweed germination.

Applying Compost

Once the bed surface has been prepared, the next step is to decide whether to apply compost. Compost can be used similarly to mulch to smother a previous crop, but we rarely do this because it would require such a large volume of compost. However, when initially shifting a tilled and weed-infested field into a no-till system, applying a layer of compost 1 to 2 inches (2.5–5 cm) thick is very helpful. A layer that thick can smother closely mown vegetation, and it certainly reduces the ability of the weed seed in the soil to germinate.

A preplant application of compost is also our means of applying minerals that are in general need over the whole crop production area (incorporating minerals in the composting process is explained in detail in chapter 10). In 2018, for example, about 75 percent of the beds at Tobacco Road Farm received at least a small amount of compost before seeding or planting. Usually we spread a layer so thin that the soil surface is still visible through the compost. The decision to apply preplant compost depends on many factors, including how well the soil biology is functioning and how much nutrient the crop will require. If there are plenty of residues and mulches still in place and we're seeding a crop

FIGURE 4.9. When we use wheelbarrows to dump piles of compost, we leave the small piles sitting until right before seeding to keep the compost from excessive drying on a sunny day.

that has low demand for growth nutrients, such as carrots, we may not apply any compost. If food sources for soil biology are lacking, however, and we're going to plant a heavy-feeding crop such as cabbage, we spread a ½- to 1-inch (1.3–2.5 cm) compost layer on the bed surface. The application of compost and mulch on the soil surface year after year has very effectively buried weed seed left from our tillage years. Compost application also aids in better seed-to-soil contact when we are broadcast seeding. In addition, the compost assists by supplying some readily available nutrients for young plants.

We usually apply compost to our 36-inch (1 m) beds using wheelbarrows, which we load using an International 240 tractor. We line up three wheelbarrows side by side and dump compost from the bucket of the tractor into the barrows. Then we roll the barrows down the wheel tracks and dump spread on the bed surface. This is surprisingly quick and effective. We have a Farmall Cub with a rear-mounted hydraulic dump cart that we can load up with compost as well, but maneuvering the tractor onto the bed surface and then turning around and reloading has proved a slower process than spreading by wheelbarrows. We don't use the 240 to spread compost directly because it does not fit

FIGURE 4.10. No tilling required! These beds with compost spread and raked out are ready for seeding, after which we'll reapply mulch.

the wheel tracks for direct delivery and is a heavier machine than we want to run over the earthworm-filled field.

On the wider beds bordered by 10-inch (25 cm) wide wheel tracks, we use a dump truck with single rear tires to dump spread compost. The truck's tires fit the bedding pattern well, and the hydraulics are of great assistance for a job like this. Frequently, two people with manure forks follow behind the truck, tossing compost onto the beds on either side of the one driven over. This manner of spreading on three beds at a time saves trips over the field, which avoids unnecessary compaction by the truck tires. We also have a manure spreader pulled by a Farmall Super C that fits over this bedding pattern. The spreader has its rear paddles disengaged and a limiter board installed to spread compost directly onto the bed below and not to the side. We utilize the manure spreader if the compost pile is near the field where we want to spread it, but the dump truck if the compost needs to be moved to more distant field locations.

When using wheelbarrows, we leave the piles of dumped compost sitting until we are ready to seed or transplant. This is significant because compost is very prone to drying out and thus being biologically damaged. We limit ourselves to dumping compost only on beds that we know we can plant the same day, and all the better if it's in the early morning or toward evening; better still if it's a cloudy or rainy day. Wide bedding rakes are the tool of choice to spread the compost across the bed surface, breaking up clumps as we go. Then it's on to seeding or planting.

CHAPTER 5

Seeding and Transplanting

Seeds are a continuation of a plant's development, and thus it is beneficial to start with seed from well-raised plants. Seeds represent the ability of the previous generations of plants to adapt to their surroundings. Seed quality is directly determined by how well a plant was able to grow in its environment, combined with how carefully it was treated at harvest and processing and how it was stored.

Considering that seed is a continuation of the previous generation, do growers want to continue the growth of small, nutrient-deficient, insect- and disease-ridden plants? Instead, if possible, it is best to continue the growth of robust, vital, insect- and disease-resistant plants that produced large, well-formed, heavy seeds. The continuation of disease through the seed is often referred to as seed-borne disease, but if the seed was well grown then there is no disease to be borne. It is certainly possible, though challenging, to start with poorly grown seed, even seed ridden with seed-borne disease, introduce it to a very healthful growing environment, and have success. However, it is best to avoid this situation and begin with healthy seed.

Seed Quality and Germination

When assessing seed quality, two simple indicators are useful: weight and size. The weight of seeds is highly reflective of quality characteristics. The more seeds it takes to make a pound, the smaller and/or lighter those individual seeds must be. Seed dealers often list number of seeds per pound, and this is well worth paying attention to, though it may not be as accurate a quality indicator as weight per *volume*. This can be measured by comparing weight of seeds with a volume, as in pounds per bushel, which gives the grower a way to measure the density of seed. The number of seeds per pound has been dramatically rising in recent years on many vegetable seed varieties and certainly indicates lower availability of high-quality seed. This is very concerning, because early seedling growth is tied directly to seed quality, and early seedling growth has very strong correlations with crop quality and yield.

Seed size is also a reasonable indicator of quality, though possibly a little less accurate. When examining seed for quality, it should be large, undamaged, and plump. Proper processing and storage are also essential to seed quality. As this can be hard to assess visually, it is often best to purchase seed from well-regarded seed companies that have invested in appropriate storage facilities. High-quality seed is often hard to find in the commercial marketplace. Because of this, growers can greatly benefit by growing and saving seeds themselves, thereby assuring a supply of the highest-quality seeds most adapted to their unique environment. However, for most growers of large vegetable diversity, there are probably limits to how many types of seeds they can raise and save.

As much as seed quality is important for a crop, so is the soil that the seed is put into. A well-balanced, fertile soil will aid the germination and growth of a young seedling, and even assist in overcoming any potential shortcomings. On the other hand a damaged, poorly cared for soil will exacerbate seed difficulties and add to the stress of the young seedling. Much crop potential is gained or lost in the very early stages of growth; this is yet another reason that having a fertile soil in place is the grower's most rewarded endeavor.

When considering the seeding rates for vegetable crops, seed quality is one of several important factors to consider. Other factors include germination rate as stated on the seed package or determined by germination tests on the farm, soil temperature, weather conditions, and the crop density best for growth conditions. When germination rates are low, it generally means the seed is old or of poor quality so an appropriate increase in seed rate is in order, often in excess of what the package states. (Replacing the seed with better quality is a better option if possible.) If germination rate of seed is questionable, try testing the seed by placing a hundred seeds between the layers of a damp towel, and place the towel in an environment of appropriate temperature for germination of that crop. Keep the towel damp for an appropriate number of days and then count the seeds that sprout, giving a germination percentage (%). Such testing is often done in winter when growers have more time. Since this is well before anticipated planting dates, replacement seed can be purchased in a timely manner if necessary.

Soil temperature is a major factor to consider when deciding on seeding rates because of its impact on germination potential. The warmth-loving crops such as beans and sweet corn will not germinate well in cold soils; often the seeds will rot if sown too early. What few plants do manage to germinate and grow will not produce as well as later-seeded ones. On the other hand, many cold-loving crops including lettuce, spinach, and mâche will not germinate in soil that is too warm. Again, although some seed may eventually germinate and grow, the plants generally do not produce the full stands that later, more appropriately timed sowings do. It can be helpful to take temperature readings with a soil thermometer to associate with seeding rates, and we do monitor this. However, a primary method we utilize is to observe the growth of weeds of a similar nature to the crop. For example, when frost-intolerant weeds are germinating in the spring, we begin seeding of warmth-loving crops. When the winter-hardy chickweed begins germinating in late summer, we start the spinach and mâche sowings.

Many of the techniques described in this manual have the effect of warming soils in the colder months and cooling soils in the warmer months, so again, there are many accompanying benefits from careful stewardship of the soil. If soils need additional warmth, they can be demulched, blackened by covering with organic materials, or covered with plastics. Employing low and high tunnels warms the soil as well. If soils need cooling they can be mulched or shaded with row covers or other barriers to the sun.

A useful technique to promote germination of seed in soils that are too warm is to put the seed in the freezer. Then in the evening, after the sun is off the field, seed the crop, and irrigate it immediately with cold water. This technique works well for summer seedings of lettuces as well as spinach. Another option is to presoak the seed in an appropriate temperature environment to stimulate germination, such as a root cellar, and then sow the seed as soon as germination is noted. Soaked seed is a little harder to work with because the seeds tend to clump and stick to surfaces when sowing, but despite this stickiness, this method can be worthwhile not only for getting crops to germinate in too-warm soils, but also to speed up slow-germinating crops such as parsley.

Direct Seeding

The decision whether to direct seed or transplant a particular crop is an individual decision for each grower depending on their situation and soils. In cases of damaged soils or intensive weed pressure, transplanting can be preferable. But in a no-till system as soils improve and weeds become scarce, direct seeding becomes a more efficient choice for many crops. There are basically two methods of direct seeding: broadcasting seed onto the bed surface or seeding in rows. When seeding in rows, a furrow is usually opened and seed is placed in the furrow and then covered over with soil. Table 5.1 provides a crop-by-crop overview of our seeding methods, seeding rates, and transplant spacing.

The primary advantage to seeding in rows is that it allows for weed control using hoes or other cultivation tools. Other potential benefits for seeding in rows are accurate, even spacing; sufficient coverage of large seeds with soil; and

TABLE 5.1. Planting Rates 2019

Crops	Planting Method	Broadcast Seeding Rate (fluid ounces / ml)* or Seeder Plate	Spacing (inches/cm)
Arugula	broadcast	⅛–⅙ (4–5)	
Sylvetta (wild arugula)	broadcast	1/20 (1.5)	
Basil	broadcast	⅙ (5)	
Beans	EarthWay seeder	Plate 14 (× 2)	22 (56) between rows
Beets	broadcast or EarthWay seeder	¼–½ (7–15) or Plate 5 with holes enlarged by drilling	15 (38) between rows
Borage	broadcast	½ (15)	
Broccoli	seedbed sown, then transplant	⅙ (5)	12 × 22 (30 × 56)
Brussels sprouts	transplants		18 × 22 (46 × 56)
Cabbage	seedbed sown, then transplant	⅙ (5)	12 × 22 (30 × 56)
Callaloo	broadcast	1/16 (2)	
Carrot	broadcast	⅙ (5); more in summer	
Cauliflower	seedbed sown, then transplant	⅙ (5)	18 × 22 (46 × 56)
Celery	transplant		6 × 22 (15 × 56)
Celeriac	transplant		6 × 22 (15 × 56)
Chard	broadcast	½ (15)	
Chervil	broadcast	⅛ (4)	
Chinese cabbage	broadcast	1/24 (1)	
Cilantro	broadcast	1 (30) summer; ¾ (22) winter	
Claytonia	broadcast	1/16 (2)	
Collard	broadcast or transplant	⅛ (4)	6 × 22 (15 × 56)
Corn	EarthWay seeder	Plate 4 or hand-sow in furrow	thin to 6 (15) in row; 22 (56) between rows
Cress, curly	broadcast	1 (30)	
Cress, broad	broadcast	¼ (7)	
Cucumber	EarthWay seeder or transplant	Plate 26 or hand-sow in furrow	12 × 88 (30 × 225)
Daikon	broadcast	1/10 (3)	
Dandelion	broadcast	1/20 (1.5)	
Dill	broadcast	1 (30) or less	

* See the note on page 82 for details on seeding rates.

TABLE 5.1 (*continued*)

Crops	Planting Method	Broadcast Seeding Rate (fluid ounces / ml)* or Seeder Plate	Spacing (inches/cm)
Eggplant	transplant		18 × 44 (46 × 112)
Fava	hand-sow in furrow		22 (56) between rows
Fennel, leaf	broadcast	1 (30)	
Fennel, bulb	broadcast or transplant	¼ (7)	6 × 22 (15 × 56) in row
Garlic	set cloves		6 × 8 (15 × 20)
Kale	broadcast or transplant	⅙ (5)–⅛ (4)	6 × 22 (15 × 56)
Kohlrabi	broadcast or transplant	1/16 (2)	6 × 22 (15 × 56)
Leeks	broadcast, or seedbed sow to transplant	⅕ (6); ½ (15) for seedbed	6 × 22 (15 × 56)
Lettuces, salad mix	broadcast	⅛ (4)	
Lettuce, head	transplant		12 × 15 (30 × 38)
Mâche	broadcast	¾ (22)	
Melons	EarthWay seeder or transplant	Plate 26 (× 2) or hand-sow in furrow	12 × 88 (30 × 225)
Mizuna	broadcast	⅙ (5)	
Mustard	broadcast	⅛ (4)	
Nasturtium	broadcast	1½ (44)	
Okra	hand-sow in furrow		thin to 6 × 44 (30 × 112)
Onion	sow in seedbed to transplant	½ (15)–⅓ (10)	6 × 8 (15 × 20)
Pak choi, baby	broadcast	1/16 (2)	
Pak choi, full size	broadcast	1/24 (1)	
Parsley	broadcast or transplant	⅙ (5)–⅛ (4)	6 × 15 (15 × 38)
Parsnip	broadcast	¼ (7)	
Peas	EarthWay seeder	Plate 14 (× 2)	44 (112) between rows
Peppers	transplant		12 × 44 (30 × 112)
Potatoes	seed pieces		12 × 34 (30 × 86)
Pumpkin	EarthWay seeder	Plate 27 or hand-sow in furrow	thin to 18 × 88 (46 × 235)

* See the note on page 82 for details on seeding rates.

TABLE 5.1 (*continued*)

Crops	Planting Method	Broadcast Seeding Rate (fluid ounces / ml)* or Seeder Plate	Spacing (inches/cm)
Purslane	broadcast	¼ (7) or less	
Radish, small red	broadcast	¼ (7)	
Radish, winter	broadcast	¹⁄₁₀ (3)	
Rutabaga	broadcast	¹⁄₃₂ (1)	
Scallion	broadcast	½ (15)	
Shallot	bulbs		6 × 6 (15 × 15)
Spinach	broadcast	⅓ (10)	
Squash, summer	EarthWay seeder or transplant	Plate 27 for large seed; Plate 22 (× 2) for small seed; or hand-sow in furrow	12 × 88 (30 × 235)
Squash, winter	EarthWay seeder	Plate 27 for large seed; Plate 22 (× 2) for small seed; or hand-sow in furrow	18 × 88 (46 × 235)
Tatsoi	broadcast	⅙ (5)	
Tokyo bekana	broadcast	⅙ (5)	
Tomato	transplant		24 × 88 (60 × 235)
Turnip, Hakurei	broadcast	¹⁄₁₆ (2)	
Turnip, regular	broadcast	¹⁄₃₂ (1)	

* Seeding rates are given in fluid ounces, which is a volume measurement, not a weight measurement. There are 2 tablespoons (6 teaspoons) in 1 fluid ounce. Amounts listed here are enough to broadcast seed over 30 square feet of growing area. Seeding rates will vary depending on many factors: weather, moon position, temperatures, compost application, seed size, germination rate, method for covering seed, surface residues, density of stand required, type of mulch. EarthWay seed plate numbers are the last numbers stamped on the individual plate. Sometimes when using the EarthWay seeder, we make two passes over a furrow. Here, this is indicated by "× 2." Unless otherwise noted, transplants are started in soil blocks in the greenhouse.

ease of placement along trellising. Sweet corn is an example of a crop that does well seeded in rows, due to the need to thin seedlings to an accurate spacing and to provide good depth of soil cover for strong germination. Pole beans, also large-seeded, do well in rows so they can be in proximity to a trellis. Crops that do well broadcast are many of the cover crops, leafy greens, and root vegetables.

BROADCAST SEEDING

Broadcast seeding can be very rewarding, and is a particular advantage of an effective no-till system where there is limited weed difficulty. Instead of being packed into rows, crowding one another for root and leaf space, the plants can naturally spread out and take full advantage of the growing area, which leads to high yield.

Broadcast seeding eliminates the need for many tasks associated with seed-bed preparation and allows for greater flexibility with associated efficiency improvements. The bed can be in rough condition with mulches, residues, and coarse compost on the surface. The lack of need for bed surface preparations is very beneficial in terms of efficiency. Also, seedbed preparation is frequently the aim of tillage. If this is not required then growers avoid both the effort of tillage as well as the associated damage to soil structure and biology. In addition, since the soil surface remains constantly covered by mulch materials, there is less need to apply mulch after seedlings emerge. On top of these benefits, broadcast seeding is very fast compared with sowing seed with a mechanical seeder.

We find many benefits from interseeding two different crops in the same growing area, and broadcasting is often the best way to interseed one crop into another. (In many cases using a seeder to intersow would be impossible.) Interseeding provides constant living plant coverage for the soil, because as one crop is harvested out, the other is already growing underneath it. It can be partic-ularly useful to interseed cover crops into the late-summer and fall vegetables.

It is generally not useful to broadcast seeds in growing areas that are prone to weed growth, because weed control options will be dramatically limited. Without defined crop rows, hoes and many tractor-mounted cultivators cannot be used effectively, and the only choice will be to pull out individual weeds by hand or cut them off with weed knives. This is fine if there is only an occasional weed, but if many weeds are present it becomes inefficient and seeding in rows may be most appropriate.

The general approach to broadcast seeding rates is to sow a prescribed weight or volume of seed to a given area. Cover crop seeding recommendations are often given as pounds per acre, but for vegetable seeds, a more meaningful rate is fluid ounces per square foot. We find an increment of 30 square feet useful. This is equivalent to 10 bed feet (3 m) on our 36-inch (1 m) wide beds, or 6 bed feet (2 m) on our 58-inch (call it 60-inch, or 5-foot / 1.5 m) wide beds. Working with this small number of bed feet allows us to quickly multiply our seed need for a given area. For example, seeding lettuce in our 36-inch beds at the rate of ⅙ fluid ounce per 30 square feet (10 bed feet), we would need ⅙ × 12, or 2 fluid ounces (60 ml) of seed, to sow a 120-foot (37 m) long bed.

The grower needs to be particularly aware of variables that impact the rate or volume of broadcast seed that is required for a given area. These variables include the size of the seed, general germination rate, weather and moon conditions, soil temperature, whether compost was applied, level of surface residue, what mulches will be applied on top of the seed, plant density required, and which tools will be used for covering. When crops are sown by volume, the size of the individual seeds will have a dramatic impact on number of plants per area. The smaller the seed, the more the seeding rate may need to be reduced. Therefore attention needs to be paid to seed size. Aside from being visually obvious, seed size is generally reflected in the seeds per pound number, which should be stated on the seed package. In general, seed size from commercial offerings of vegetable seed is trending toward smaller seeds. On top of that we've noticed that seed sizes of certain crops are highly variable, both from batch to batch of a variety and among different varieties. These crops include beets, chard, carrots, and spinach.

Applying compost to the soil surface has a major influence on broadcast seeding rates. Seed mixed into a compost provides excellent germination potential, so often rates can be reduced. Seed applied to a mulched or (crop) residue-rich environment, by contrast, may not be exposed to appropriate moisture for germination, especially if the residues are thick. This would call for increased seeding rates. The tools used to incorporate the seed, such as rollers, drags, or rakes, also impact the rate. Generally the more thorough this step, the lower the rate can be. A final point of consideration is the type of mulch applied over the broadcast seed. If no mulch will be applied, increase the seed rate. If a chopped or partially decomposed leaf or wood chip mulch is used, the chart rates listed in table 5.1 may be best. The broadcast seed rates on the chart are the volumes that we use for beds that are composted, dragged, and mulched over. If the mulch will be straw, then increasing the seeding rate a bit is advisable.

We use two sizes of measuring cups for determining volumes of most broadcast vegetable seed. One cup (a small plastic "medicine cup") measures a fluid ounce in dram (⅛ fluid ounce) increments; this cup is useful for small seeds like arugula and lettuce, where a fluid ounce of such seed will generally cover 180 square feet (16.5 sq m). For greater volumes, we use an 8-fluid-ounce measuring cup. Both cups are transparent (see figure 5.1), which allows us to quickly gauge the amount of seed initially required in the cup and remaining as the seed is spread. For example, when a quarter of the given area has been sown, we know there should be three-quarters of the seed remaining in the cup. When half the area is sown, half the seed remains, and so on.

FIGURE 5.1. Measuring cups like these are useful to gauge the rate of seed that is being applied over the bed area.

FIGURE 5.2. Seed covering tools: a ring drag made from grain drill seed covering gangs, two leaf rakes with straightened tines, the Garden Weasel for extra churning action, and the roller for smoothing the bed surface and better seed-to-soil contact.

It is important to carefully scatter the seeds to achieve an even spread across the bed surface. We scatter seed by hand. This is a critical skill for a grower to develop, and it certainly is not hard to do so. One approach is to pour a given portion of the seed, say one-quarter of the total volume, into the seeding hand, and then spread this seed onto one-quarter of the total area to be seeded. We find it best to sow only half the width of a bed at a time, so one-quarter of the seed should cover half the length of half the width of the bed. Then pour the next quarter of the seed into the seeding hand, and continue. If hand-broadcasting seed is new to you, it is usually best to start off on a smaller area, then progress as you gain proficiency. Perhaps getting started with vigorous, easy-to-grow greens like arugula or cutting lettuces is advisable. Once you develop your technique, seeding in this manner can be very fast, and there is great ease in switching from one variety of seed to another.

To encourage germination and get ahead of any weed germination, work the seed into the bed surface right after sowing. Useful tools for this task include drags, rolling cultivators, and roller/packers. For our scale of 3 acres (1.2 ha) of production, we primarily rely on hand-pulled versions of the drags and rolling cultivator, and even pulling the roller in some instances. When we need to roll a large area, we generally use a tractor-mounted roller. There are many ways to fashion a drag, as shown in figure 5.2.

or otherwise ganged seeders can deliver accurate between-row spacing, but many growers choose to rely on hand-pushed single-row seeders due to their ease of use. One way to accurately space furrows when using a hand-pushed single-row seeder is to lightly premark the furrows using a tractor mounted with shovels or tines on the cultivation frame. This is particularly useful if tractor-mounted weed control equipment will later be utilized. It is also useful in a high-residue environment, because the cultivating shovels or tines will clear residues as they mark the furrows, which will make seeding easier with a seeder. A hoe can also do the job of opening a furrow for an easy pass with a seeder when a lot of residue is present. Our favorite hoes for this are the Warren hoe or other plowlike hoes. Another row premarking tool is a wide rake with plastic tubing pushed onto the appropriate teeth and then dragged down the bed. Whatever tools are used to make the furrows, it is preferable to seed them immediately to preserve moisture.

Once the furrow is marked or opened up, the next step is to place seed by pushing a seeder down the row. Our favorite for this task is the EarthWay seeder. The EarthWay has a V-shaped furrow opener, a depth-adjustable seed chute, a chain drag to cover the seed, and a rear packer wheel. It is relatively easy to operate in our fairly high-residue environment, especially with a little pre-opening of the furrow as described above. A rotating plate in the seed hopper picks up seed and delivers it to the chute. We have purchased many extra plates, and in order to achieve more accuracy in seed spacing, we drill out the holes in the plates in increments to accept more or slightly larger seeds. We also caulk over some of the holes in varying degrees in order to achieve wider spacings. This refinement allows us to use the EarthWay for very effective and accurate results. Unlike many other seeders, it is easy to fill and empty the seed hopper. When pushing the seeder down the bed, it's best to hold it at a slight angle so that it "leans" toward the seed plate. If there are obstructions in the seed furrow, the seeder can simply be lifted and placed right back down on the far side of the obstruction.

One design flaw with this seeder is that small round seeds can wedge behind the seed plate, which lifts the plate away from the side wall and thus fails to deliver seeds to the chute. To avoid this, we simply refrain from filling the hopper with the seeds, as seeds in front of the plate are the ones that get wedged. This may require that we carry the seed packet with us for more frequent refilling than usual. Another tip is that if you are seeding on a slope, push the seeder uphill rather than down.

Other makes of seeders may be as effective as the EarthWay, and for different situations various seeders may be more appropriate. We now plant most of our crops by broadcast seeding and transplanting, but we still use the EarthWay to plant beet, chard, and winter squash seeds, along with a few other crops.

FIGURE 5.4. Cultivator shovels for furrowing mounted on the tractor toolbar allow for fast work where long furrows are needed.

FIGURE 5.5. Furrow and seeding tools. The EarthWay seeder (*left*) and several hand tools: a plow-shaped furrowing hoe, our hoe of choice for furrows; swan neck hoe; single-tine ripping hoe, to loosen the furrow bottom if needed; old-style Warren hoe; and two long-bladed trenching shovels, which are used for removing soil to set transplants.

FIGURE 5.6. A trial of different seeding methods shows excellent stands of arugula on a finely raked bed seeded with the pinpoint seeder (*left*) and on a bed where seed was broadcast and churned into the soil surface with a drag (*right*), which was much less work raking the bed. An unprepared bed seeded with the EarthWay seeder (*center*) has a few skips, but the overall stand is still fine.

Even at our scale of production, we still choose to place some seed by hand. Since we often have to pre-furrow in high-residue beds anyway, it is easy enough to then seed sweet corn, peas, squash, melons, cucumber, okra, and beans by hand, as long as we are seeding only a few hundred feet. Careful hand-seeding and covering with a hoe usually result in a better-spaced, more consistent stand with these large-seeded crops than does sowing them with the seeder. In our tillage days, we made use of a four-row pinpoint seeder, but it sits abandoned now. It was a useful implement when our beds were well tilled, firmed, and shaped, but that is simply not what the beds look like anymore, and to bring them to that condition would not be efficient.

When crops are sown at the appropriate rate, little to no thinning will be necessary. As a rule, crops that require very accurate placement are transplanted, but a few crops that are commonly direct seeded usually require thinning: squash, cucumber, melon, corn, and okra. These fruits of summer are relatively intolerant of crowding and do not transplant particularly well. These crops are all easy to thin, and thinning presents an opportunity to select the best plants. An important consideration when thinning: It is better to leave vigorous plants that may be closely spaced and thin out weaker ones than to try to achieve a perfect in-row spacing. For example, to seed 100 row feet (30 m) of cucumber that requires an in-row spacing of 12 inches (30 cm), we will try to leave in place the most vigorous 100 plants in that row overall. Of course this is within reasonable limits. There need to be enough appropriately placed plants to eventually cover the allotted growing area.

Transplanting

Setting transplants assures a full bed of plants all at the correct spacing, gives the grower a jump on any weed difficulties, and can increase the rate of production of a land area because it reduces the time to harvest versus direct seeding. Another advantage of raising transplants is that sowing seed in well-made potting soil may offer the seed a better start when the soil environment in the fields is still being improved and may be imbalanced. The major drawback of transplanting is the labor required.

We choose to transplant umbelliferous crops such as celery that, because of slow germination, are difficult to direct seed, crops that need to start life in a hot environment (such as tomatoes and some squash-family plants), and those that require very careful spacing for proper yield, including head lettuce and brassicas such as cabbage and cauliflower. When raising transplants, growers concentrate the care of large amounts of small plants into a limited area. These plants will propagate much larger areas in the field. When considering how to concentrate these seedlings, growers have two options: growing in flats with potting soil or growing transplants in the field in seedbeds.

Seedbeds are concentrated growing areas in the field often with very high-quality soil and protective structures such as cold frames, low tunnels, or high tunnels in place. They are very useful for growing large amounts of seedlings with potentially minimal care compared with raising plants in flats. We utilize seedbeds extensively for brassicas and alliums. These crops all transplant well from the essentially bare-root transplants that seedbeds yield. Seedbeds are particularly useful for starting seedlings in seasons when it is too hot or cold to maintain production in flats efficiently. Examples include fall-sown onions and leeks grown in unheated high or low tunnels through winter and June seedbed production of cabbage, broccoli, cauliflower, and kale seedlings that will provide the fall harvest.

Potting soil culture provides reliable results for crops that do not transplant well (umbels and early cucurbits) or that need very warm conditions and more care in their early stage (solanaceous crops). For crops that do not transplant well, starting the seedlings in cell trays, pots, or soil blocks yields transplants that can be set with a minimum of root disturbance. Other crops like alliums can be raised in flats filled with potting soil and then set in the field by separating into mostly bare-root seedlings.

Because we direct seed so many crops or raise the seedlings in seedbeds, we use 2-inch (5 cm) soil blocks for starting the limited number of transplants that

FIGURE 5.7. The leek plants in this seedbed, which was sown five months prior in early December, are ready for transplanting.

do best growing in potting soil (for details on how the potting soil is made, see chapter 10). Soil blocking is a superior method for producing very high-quality transplants. The roots are well placed, and there is a relatively generous amount of soil for small seedlings. The blocks transplant well into the field with minimal root disturbance or transplant shock. Almost all transplants survive, so there is no time wasted on resetting beds. The growth of these crops is steady with no stunting. This ability to maintain steady growth can dramatically improve later stages of growth.

Transplants growing in soil blocks are set in the field directly as 2-inch blocks except for the solanaceous crops, which are potted up from 2-inch blocks to 5-inch (13 cm) pots. The soil blocks all receive two seeds to start, which allows us to thin to the better plant and results in almost 100 percent filled blocks. The soil blocks sit in flats on benches in minimally heated hoop tunnels from February to June. Later soil-blocked transplants, if required, are grown outside on benches under low tunnels or other structures covered with shade cloth or row cover. The umbel seedlings do much better under a cloth row cover low tunnel or other support structure inside the greenhouse during the February-to-June period as well, because of their preference for some shade.

Careful watering of seedlings is important; we use either a fine-spray Wonder Waterer or a watering can with a fine rose end. The warm water is supplied to the watering wand by mixing the hot and cold water lines at the house system outputs with a Y hose attachment. This needs to be shut off after use or the hot and cold water may try to cycle inappropriately. As a reservoir for filling watering cans, we keep about twenty 5-gallon (20 L) buckets of water in the hoop tunnel that heat up with the sun. A small amount of nutrient is supplied by mixing a bit of liquid seaweed, molasses, and raw milk into the water in the watering cans. It is very important, however, not to overwater or overfertilize seedlings.

Transplants are set in the field at a relatively young age to lessen transplant shock and provide for consistent growth. Evenings and mornings are the most appropriate times for setting; cloudy, rainy days are even better. Transplants are set quickly and roots are not allowed to dry. We produce seedlings in excess so that we

FIGURE 5.8. These 2-inch soil blocks can grow impressive seedlings. With practice, 100 blocks from our farm-generated potting soil can be made in approximately five minutes.

can select only the best of them to set in the field. For bare-root transplants or 2-inch block transplants, we rip furrows with shovels mounted on the belly cultivator frame of the tractor or by hand with furrowing hoes. Bare-root transplants are dug from their seedbeds with trowels and put on flats for immediate transplanting. We cover the roots in mulch and water them while they wait. Soil blocks can be transplanted right from their trays. We aim to set and firm a transplant at least every 6 seconds, hopefully even faster; a tray of 100 plants takes 5 to 10 minutes to set. It is important to firm in the transplants with soil, not with undecomposed mulch or residues. When setting any transplants into the field, it is best for soil quality and weed control to create as little disturbance as possible. Transplanting through thick organic mulches requires a little finesse. Use a trowel or narrow planting spade to poke holes and remove some soil. Then set the transplant in place and tuck it in with the removed soil. Developing this skill is worth the effort since there will be few or no future weed control needs. Often compost can be laid down preplant on top of the mown residue and underneath the added mulch materials. If this is not

FIGURE 5.9. Umbelliferous seedlings in flats grow under a row cover shade cloth.

FIGURE 5.10. Kale transplants set in spring under cloth row covers. The mixed mulch surface characteristic of our no-till system shows well here.

possible, crops will be fertilized by the decomposing mulch materials combined with side-dressing, often liquid in this case. This method may work best with large transplants such as tomato, pepper, and eggplant, though we have set smaller transplants like cabbage as well.

To set large transplants in 5-inch pots, we dig individual planting holes using a trenching shovel instead of furrowing. The handle is marked in increments of inches so that proper spacing can be determined simply by laying down the shovel on the bed surface. There are also markings on furrowing hoe handles so that the furrow spacings can be quickly checked. All seedlings are set a little bit lower than ground level and the holes are filled in with soil, again avoiding undecomposed materials. All transplanted seedlings are then immediately watered in; they will also be remulched if necessary, particularly if a significant amount of soil has been exposed to the surface. We often set up low tunnels covered with cloth row covers over transplants, particularly if they were taken directly out of the greenhouse environment to the field and not first hardened off outside.

Often the whole process of transplanting is a steady workflow, with one person ripping furrows while other people are digging or collecting the transplants and delivering them to still others who are transplanting into the furrows. As well, a person follows after the planters, watering in the seedlings already set. A similar process is followed when planting potatoes and garlic.

Proper labeling of sown or transplanted crops is very important. Labels are cut from leftover lumber and can be up to several feet long to mark tall, dense crops like tomatoes. Each label states the variety name, the date of seeding and possibly transplanting, the seed rate utilized, and any other pertinent information such as variances in covering method or fertilizer application.

The early stages of crop growth are highly determinative of the end result, so it pays for the grower to give significant attention to seeding and transplanting. This is similar to setting the stage well with efforts to soil fertility. The work and effort done before and during these early stages saves much effort that would otherwise be expended later in the growth cycle to make up for deficiencies.

CHAPTER 6

Mulching and Irrigation

The use of mulch and irrigation are of significant benefit to no-till vegetable systems if handled appropriately. Both are essentially fertilizer materials that can have dramatic influence on crop growth and balance. The selection of mulch materials, the quality of water for irrigation, and the volumes and timing of application all require grower consideration. In our no-till system the soil is always covered with a mulch material selected for its influence on the growing system. Irrigation is provided whenever seed is sown or transplants set in the field. Irrigation also is a primary means to apply side-dressed liquid fertilizer as the crop is growing. With the improvements in soil characteristics from no-till, fertility balance, and mulching, irrigation is rarely required to keep the soil hydrated. This has been a significant savings in labor for crop production. Though on some occasions we have applied mulches before a crop is transplanted or sown, in general we mulch and irrigate immediately afterward.

The soil surface is where the four elements of sun (warmth), air, water, and earth all meet and interact. How the soil is covered has dramatic impact on that interaction. Covering soil with a decomposable material creates a functioning natural cycle of life and decay. A surface mulch provides for the decay and release of nutrients from the previously living organisms, and it also protects the soil from the drying effect of the sun and wind, as well as maintaining a steady soil temperature and moisture supply and keeping the soil surface from eroding under heavy rain. All of this results in a hospitable environment for the appropriate soil life with all the benefits derived from this in terms of soil development. Mulch usage in vegetable production increases growth and yield of crops, and provides excellent weed control as well. Mulch makes for a clean harvest because soil is unlikely to splash up onto leaves and fruits when the soil surface is well mulched. The improved soil aggregation under mulches yields root vegetables that clean up easily.

It is useful to look at composting and mulch application as a continuum. A main objective of applying compost, mulch, and other fertilizers is to enhance the growth and functioning of soil life in order to provide for abundant crop growth. Selection of mulch materials is thus part of the overall fertility plan. Whether mulch is a prepared compost or some other organic material, it does serve as a fertilizer, dramatically impacting the carbon to nitrogen balance in the field, as well as influencing other nutrient levels, bacterial:fungal ratios, and soil air and moisture.

Mulch Materials

Most mulch materials are of an organic (carbon-containing) nature with straw, leaves, and woodchips being common examples. Non-organic mulch materials include minerals like stones and rock dust and various plastic sheeting materials. Though we utilize plastic sheeting for solarization at Tobacco Road Farm, we do not apply it directly upon the ground as a mulch around growing plants. We find that mineral and organic mulches help us produce better soils and better crops. There are several downsides to black plastic mulch. For starters, covering soil with plastics long-term can cause overheating and have a detrimental impact on soil air and moisture. They preclude the possibility of applying decomposable materials to the soil surface while in place, and limit the area available to maximize soil coverage with growing plants. If soils or crops would benefit from more heat, we place low or high tunnels over the bed. This suspends the plastic in the air and thus avoids the problems related to plastic-covered soil. Another drawback of plastic mulches is that they must be picked up and disposed of at a loss, whereas organic mulches decompose and continue contributing benefits for years to come.

Most of the materials we use as mulch are also raw materials for the compost we make. (The details of how to handle and procure materials for mulch and composting, as well as contamination issues, are examined in chapter 10.) In our system a frequent mulch material is the chopped residues of the previous crop or cover crop that grew in the area. This is particularly the case in the spring when chopping overwintered cover crops generates large volumes of material. Thus to some degree mulch materials are produced right in place where they are applied. However, it is often beneficial to bring in additional materials for mulching. Some common ones are leaves, straw, wood chips, hay, green grass, sawdust, bark, coffee grounds, and cardboard, as well as crushed stones.

LEAVES

Fallen deciduous tree leaves are very abundant in our environment and are the bulk of the natural mulch material that sustains the forests in our region. As such they are generally well suited to our soils; they are our primary material. We often mix leaves with some straw, wood chips, and crushed stones. Earthworms rapidly assimilate leaves into the soil, especially when the leaves are pre-chopped or partially decomposed. The remaining straw, wood chips and crushed stones persist, offering soil coverage for longer-season crops.

Leaves are much easier to handle and generally more effective when partially chopped; fortunately this happens when leaves are collected by landscapers' vacuum systems. There often is some coniferous leaf (needle) material mixed into the leaves, and diversity here seems to be of benefit. Oak leaves in particular may be helpful in repressing slugs. Leaves are generally carbon-rich, though less so than straw or wood materials.

STRAW

Straw for mulching can sometimes be grown in place in the form of overwintered cover crops from the grain family, usually chopped or mown in the spring, although other timings are possible. Cover crops grown for mulches are best blended for appropriate C:N ratio as well as kept weed-free. Purchased straw often contains grain seed that can turn into aggressive weeds. To prevent this, straw can be laid out, and then rain allowed to fall on it, which will cause the grain seed to sprout. When laying out straw, it can be spread out by hand, or bales can be run through a bale chopper, which improves its ability to be applied evenly to the field surface. When the straw is later moved for usage, the sprouted seedlings are killed. Allowing straw to sit outside in the rain and age a little bit is also beneficial because it prepares it for decomposition by soil biology more easily. Straw is relatively carbonaceous, and fungi are particularly fond of straw when it is applied to the soil surface. Straight straw mulch often does not result in as crop growth as vigorous as does leaf or wood chip mulch, but it performs well in mulch mixtures.

FIGURE 6.1. In this trial of different types of mulch, the straight straw materials (lighter areas) did not produce as consistent germination and growth as did the wood chips, leaves, or mixture in the darker areas of the bed.

WOOD MULCHES

Wood chips are often combined with leaves, depending upon the time of year when the material was chipped. The chips have excellent durability on the soil surface and can provide cover for longer periods of time than straw or leaves. Wood chips are more carbonaceous than these other mulches, and their use may not be appropriate for soils that have high C:N ratio. Fungus does favor wood chips as well as straw. Wood chips can be a very useful mulch for tilled, high-nitrogen, bacterial-dominated soils as they can begin to restore balance. Chips particularly encourage fungal growth, and the high carbon levels help balance the high nitrogen of the soil. Chips from land-clearing projects tend to be more carbonaceous than chips produced from roadside branch removal.

Bark mulch is similar to wood chips in many regards, although it is a material of exclusively exterior origins while wood chips are a mix of exterior and interior. Sawdusts generally have very high carbon content and may break down more slowly depending on treatment, as described in chapter 10. Allowing sawdusts to age can improve their characteristics for mulch usage, but weed seeds must be kept off the piles as they sit.

HAY

Hay is a common mulch material, though it does offer the potential problem of weed seeds of perennial nature. These can be very detrimental in a no-till system, where perennial weeds present the greatest difficulty to the continuity of non-soil disturbance. Perennial weeds do not readily die with quick

FIGURE 6.2. This mulch pile includes three of the materials we use most: straw, leaves, and wood chips.

solarization. When levels build up in fields, they can overwhelm hand removal methods, and tillage will need to be utilized to reestablish control. Running hay bales through a bale chopper produces a material that is very easy to spread, even in areas that are difficult to mulch, such as around closely spaced onions. Hay is variable in its carbon to nitrogen ratio; some legume hays are relatively nitrogen-rich whereas coarse hays of low feed quality are often more carbon-rich. Green lawn grass cuttings are a relative of hay. Lawn grass is usually high in nitrogen and may be relatively weed-seed-free depending on the time of year it is cut. Grass can be an excellent addition to mixed mulch materials in that it provides a little bit of nitrogen and other nutrients, and thus may bring better balance into the mix. Grass piled by itself heats to very high temperatures very quickly, but diluting it by mixing into other mulch materials like leaves keeps it cooler. Grass applied thinly around plants does not heat and can supply fresh nutrients to the soil life, but it does quickly dehydrate and decompose, and can leave the soil exposed unless more mulch is applied.

STONE MULCH

Stone dusts are non-organic mulch materials generally created by the crushing of rock at quarries. Rock crushing results in the creation of fragments of various sizes with the finest being referred to as stone or rock dust. These materials are variable in their base constituents as well as the fineness of grind. Being the least valuable product of rock crushing, they often are quite affordable. The rock in our Northeast region is often high in silica and therefore assists plants in exterior hardening, constricting excessive growth, and strengthening reproductive efforts. The grind of the material can dramatically influence the ease of usage for mulch. Coarsely ground rock dusts (often with small chips present) are less likely to clump and thus easier to spread evenly than finely ground dusts. Finely ground rock dusts have a claylike texture and influence on the soil, with improved nutrient release of the minerals. The fine dusts can also pose a difficulty to the lungs, because when surface applied as a mulch, the material dries and later becomes airborne if the mulch is disturbed. Because of this hazard, we incorporate fine stone dust via the composting system and use only coarse stone dusts in mulching. We run trials by applying different stone mulches independently to assess their influence on living plants and soil systems, but when mulching, we usually mix the coarse rock dust into organic materials.

COFFEE GROUNDS

Coffee grounds are similar to green grasses as a way to bring a higher nitrogen content into the mulch mix. With the amount of coffee that is consumed, there are a lot of coffee grounds around, if you seek them out! We backhaul them from our grocery and restaurant customers. Caffeine is a nitrogen-containing alkaloid and a powerful plant stimulant. Another reason why coffee grounds maybe a particularly helpful mulch additive is that coffee repels slugs. We generally mix coffee grounds into our mulch piles as they arrive. Other backhauled seed meals and various other processing leftovers can be mixed into mulch, too.

CARDBOARD

We generally don't use cardboard in the annual vegetable fields, but it is very useful in keeping perennial weeds out of the perennial herb gardens, brambles, and blueberries. We often pile additional mulch materials on top of the cardboard to hold it down. Some growers utilize cardboard in annual vegetable plantings by poking holes into the cardboard to set transplants through and then covering the cardboard with additional mulch materials. Earthworms and soil life do appear to thrive underneath the cardboard layer.

Managing and Applying Mulch

We maintain separate piles of wood chips, leaves, straw, and stone dust. The piles are kept free of weeds by piling them upon a weed-free base and removing any weeds that sprout and grow in the mulch before they go to seed. To make a mixed mulch, we use the loader tractor or, for small amounts, mix it by hand using pitchforks. The loader tractor fills the appropriate spreading equipment: dump truck, manure spreader, large plastic two-wheeled dump carts, or wheelbarrows. Occasionally we fill flexible plastic bushel containers and haul them to the field in case of particularly finicky tasks in spots where the carts and the wheelbarrow cannot fit.

The dump truck is usually the vehicle of choice when mulching the 58-inch (1.5 m) field beds; the truck can straddle the bed surface. In this manner often three beds can be quickly mulched at a time. One person drives slowly along one bed, and two people with pitchforks distribute the material off the back of the truck across that bed and the two adjacent beds. For 36-inch (1 m) beds, we load up two large plastic two-wheeled dump carts at a time. The carts are then brought out onto the field and pitchforks used to empty them out. It is important to achieve an even spread.

Mulching occurs immediately after broadcast seeding, followed by watering. This protects any compost that was applied before seeding, covers the seed further, holds in moisture for germination, and helps deter weeds. In the case of transplanting, it is important to water in the transplants first and then apply mulch. This is because a transplant does not want to wait to be watered! Immediate watering reduces transplant stress. It is also more water-efficient,

FIGURE 6.3. Loading mulch into two large dump carts. The mulch is mixed by the loader at the piles, the carts are maneuvered out onto the field, and the mulch is distributed by hand or pitchfork.

FIGURE 6.4. A freshly seeded area with a mulch layer just applied.

because water doesn't have to first penetrate the mulch. The lower volume of irrigation water is gentler on the young transplant as well.

When mulching over seed, only a light coating of mulch materials is applied. It is often better if the mulch used is chopped or otherwise broken up. For most seedings this covering just barely obscures the soil surface. Mulches can be applied much more heavily around transplants and up to several inches thick on potato and garlic beds. When spreading mulch around transplants, a fine-textured material is best, especially with closely spaced transplants such as onions. The wheel tracks between beds also receive a thick coating of mulch— this helps feed the soil life, controls weeds well, and ends up being a source of fertility for the growing area when beds are reshaped. Mulching can result in keeping soil excessively cool in spring and thus delaying growth, a common complaint about mulch usage. Mulches do have a moderating impact on soil temperature, keeping soil warmer in winter and cooler in summer. However, mulches also greatly enhance soil life, and the increase in life adds a warmth to the soil. Therefore mulched soils with a high level of biological activity generally do not suffer from slow crop growth in spring, whereas mulching a low-functioning soil may slow crop growth significantly. If soils and crops would benefit from additional warming, mulches may be raked aside in spring.

FIGURE 6.5. Wood chip, leaf, and straw mulch covers the soil of these beds of broadcast seeded lettuce. These beds have received no weed control.

Other options would be to set a low tunnel over the cropping area or to apply dark mulch materials, such as partially decomposed leaf or mulches, that will help with solar heat gain because of their color. Mulches may even be blackened by charring, such as partially burned sawdust.

Irrigation

The management of water is a critical area of the grower's influence on the natural systems. As described in chapters 1 and 2, water has a dramatic impact on our soils and crops. As growers we seek to set up conditions and systems where water can be either reduced in the growing environment or added to the growing environment when necessary. Growers use various means to reduce water in their growing environment. These include drainage ditches, subsurface drainage pipe, and raised beds. Growers also reduce water by utilizing slope and bedding layout, increasing airflow and sun exposure, improving soil structure, and setting up high/low tunnels or other plastic sheeting soil covers.

Several strategies help to increase water in the growing environment. Irrigation is a primary method, but growers can also improve soil structure and organic matter content in the soil, mulch the soil surface, keep the soil covered with growing vegetation, or provide windbreaks and more shade. Changing from a tillage system to no-till will have a significant impact on soil moisture, as already discussed. All these approaches have implications for the critical soil air:water balance as well as the overall balance of the polarities of growth for the crop.

Vegetables are a very high-value crop, and the cost of irrigation equipment is generally low, especially if water is reasonably available. Sometimes conditions are in place where irrigation is of benefit, but it is important to carefully consider when irrigation will be of most use. Growers are prone to overirrigate, and although irrigation can make up for soil water deficiency, overirrigation can be costly in terms of both the cost of labor and the potential for irrigation to be detrimental to the crop growth. Irrigation can be detrimental not only from its overusage but also from incorrect timing and from water-quality issues. As stated earlier, the soil air:water balance and its relationship with the soil life exist naturally in a fine balance. For growers to interrupt this balance with a large input of water from a source that is not rain has dramatic ramifications.

At Tobacco Road Farm, irrigation is consistently useful for germinating newly seeded crops and watering in transplants. It is also important for preserving the life in freshly spread compost. The progression from pulling surface solarizing covers and compost application to seeding/transplanting, mulching, and watering in happens very quickly. Preferably all these activities are proceeding at the

FIGURE 6.6. Watering in newly seeded beds is critical for quick, even seed germination and hydrates the compost and mulch.

same time, as a crew of people carry out their tasks in succession across the field.

Even during years considered droughty in our region, our crops have continuously thrived due to beneficial natural soil moisture, and at this point in our farm's development the post-seeding/transplanting watering is often the only irrigation a crop will receive. The soil structure's homogeneous and intact state, the high organic matter content of the soil, the mulch on the soil surface, and the relatively high water tables of the fields mean we no longer need irrigation for any purpose other than encouraging germination, supporting new transplants, and applying side-dressings of fertilizer. For these occasions an irrigation system is very beneficial. We have fully abandoned using drip irrigation, however, at a great savings of time and materials.

Vegetable-growing soils do suffer from a biological crash when allowed to dry out thoroughly, and recovery will be slow, so a steady supply of just the right amount of moisture is important for consistency of vegetable crop growth. When growers face conditions in which moisture is not sufficient to provide for the steady growth of a crop, it is probably a profitable proposition to carefully supply the necessary water through irrigation. Warmer water may be easier on the soil life than very cold water. With this in mind, if the water source is cold, it is often best to water early in the morning when soils are cooler to avoid a shocking change in soil temperature. Early-morning watering also allows for the leaves of plants to dry during the day, which helps avoid an overwatered plant condition prone to disease.

It is important to check on the progress of the irrigation water as it moves into the soil. In order to impact germinating seed, the water must penetrate the mulch layer and wet the soil in contact with the seeds, for example. When irrigating dry soil to support plant growth, the depth of penetration needs to be carefully monitored, because, as explained, overwatering can be very damaging to soil and crop. Underwatering, however, can be ineffectual. Irrigating a mulched surface for seed germination aids in the germination of the seed because the mulch retains moisture. However, if there is limited water or time, irrigating the seed before mulching can be quick and effective.

WATER QUALITY

Surface water from ponds, lakes, rivers, or streams is a common source of water for vegetable growing. Surface water has the benefit of being exposed to the sun and generally is warmer than water from other sources. It also may contain beneficial nutrients and living organisms. Unfortunately, surface water may also contain many pollutants. Investigating the source of surface water available for irrigation is worthwhile, and so is conducting trials to assess the response of plants to the water. Due to its living nature, surface water may require filtration to prevent clogging of application equipment.

Subsurface water comes from wells, either dug or drilled. Subsurface water is substantially less prone to clogging irrigation equipment than surface water. There may potentially be fewer contaminants as well. Trialing is still a wise precaution, but if only small amounts of supplemental water are necessary for germination and transplanting there would probably be limited buildup of contaminants. Water from municipal sources is hazardous to plants and soil, however. "City water" is often highly contaminated with antimicrobial chemicals. These chemicals have antimicrobial effects in the soil and also in the digestive tracts of animals (humans) who drink this water! Some improvement in water quality may be gained by allowing time for municipal water to sit and off-gas some of the more volatile chemicals before use. Growers faced with no supply other than detrimental municipal water may need to purchase thorough filtration equipment, such as a reverse osmosis system, or perhaps invest in having a pond or a dug or drilled well installed. If the groundwater level is reasonably high, it is very impressive how quickly and inexpensively an excavator can dig a pond or a well. Rainwater in many areas is presently fairly contaminated as well, but in areas where rain does not seem to adversely affect crops, a rainwater collection system could be installed.

APPLICATION METHODS

A basic setup for most irrigation systems is a suction line to the water source with a foot valve on the end to prevent backflow; a pump; a discharge line (main line), usually of a large size; and hoses branching off from the main line into the field. Often there is a fertilizer mixing tank as well, which can be metered into the discharge line or the suction line. Many metering devices, which mix materials into the discharge line, are prone to clogging when living organic materials are utilized as fertilizer. To avoid this we work with the fertilization tank setup described in chapter 9. Pumped irrigation water can be applied to growing crops using drip lines or sprinklers, or by hand using hoses mounted with nozzles. We prefer watering with hoses, and it is the only form of irrigation we use at present. We use large-diameter (1 inch / 2.5 cm) hose lines with high-output hose ends branching off from the main water supply line.

These hoses can deliver a lot of water quickly, somewhat like a fire hose. We use different hose-end nozzles on the hoses to deliver various patterns of water from fans to cones.

The cone-shaped spray pattern is best for seedbeds or transplants. The fan pattern is particularly useful for side-dressing. In our experience the adjustable fan-patterned nozzle from the Kochek company does not clog up with organic fertilizer materials. Directing the spray along the base of plants keeps water and fertilizer off crop leaves, which is important as fertilizers can "burn" leaves, and wet leaves can add to disease difficulties. We run these hoses to a maximum of 200 feet (61 m) because they can be too heavy to pull beyond that. Watering in this manner allows very precise control over the water and fertilizer application and cuts down on over- or underwatering.

In the past we used sprinkler irrigation extensively. A network of ¾-inch (1.9 cm) diameter hose lines led out to sprinkler heads. Setting up the sprinklers was time consuming, thus it also was an expense. Initially we worked with relatively wide-circle Rain Bird sprinkler leads, but eventually we opted for smaller square-pattern spray heads from the Nelson company. The small square pattern allowed for more controlled wetting of a given area, putting out more water than the Rain Birds. The square-pattern spray heads required more moving, however, and often needed to be elevated onto buckets or crates to propel the water up and over nearby crop canopies. All the hose ends had to be fitted with in-line screens to prevent the sprinkler heads from clogging. These screens needed to be checked regularly when using surface water or organic fertilization materials, because they were prone to clogging. Though we no longer use this system, we still have the hose lines and bibb valves placed in the main irrigation lines in case we need it at some future time.

We also don't miss using drip irrigation! All those tubes everywhere, always in the way, and fittings always blowing out or a tear in the line . . . We used to call it the "irritation system." Drip irrigation can be very useful for watering vegetable crops in dry conditions, and doubtless farmers in dry climates rely on it. The one benefit of drip irrigation on our farm was the ease of supplying a little bit of soluble nutrient to a crop in a steady fashion, which now requires slightly more effort, because fertilizer materials are applied through handheld hoses. Again, our drip irrigation equipment is still around, just in case we need it someday, but it has been almost 10 years since we last used it.

CHAPTER 7

Crop Rotation and Planting Cycles

As growers we make decisions and create plans for crop selection, crop timing, labor, marketing, and more. At Tobacco Road Farm the overarching plan for when, where, and how much of each crop to produce is summarized in a large chart that we update each year. This chart is our planting schedule, crop rotation plan, and plan for supply to market all in one. It also determines our transplant needs and provides the basis for estimating labor requirements. The chart is adjusted from year to year to meet the changing demands of the marketplace. We also make changes to the chart in season to adjust for vagaries of production. It offers flexibility in many regards, while still providing an overall structure. This chart has steadily grown in complexity over the years. Its refinement has allowed for less over- or underproduction and greater profitability.

We use the chart (figure 7.1) in conjunction with a crop rotation map (figure 7.2) that shows the location of all the growing areas and how crops rotate from one growing area to another over time. The chart itself shows which crops are planted in each growing area, the time of year they are planted and harvested, and the planting sequence. For the purposes of planning, we have divided our farm fields into growing areas of equal size. In our year-round production field, the unit area is 3,000 square feet (280 sq m). In our other two fields, the unit area is 10,000 square feet (930 sq m). Thus, although crops rotate from one spot to another from year to year, we can reasonably expect consistent yields, because the unit area remains consistent.

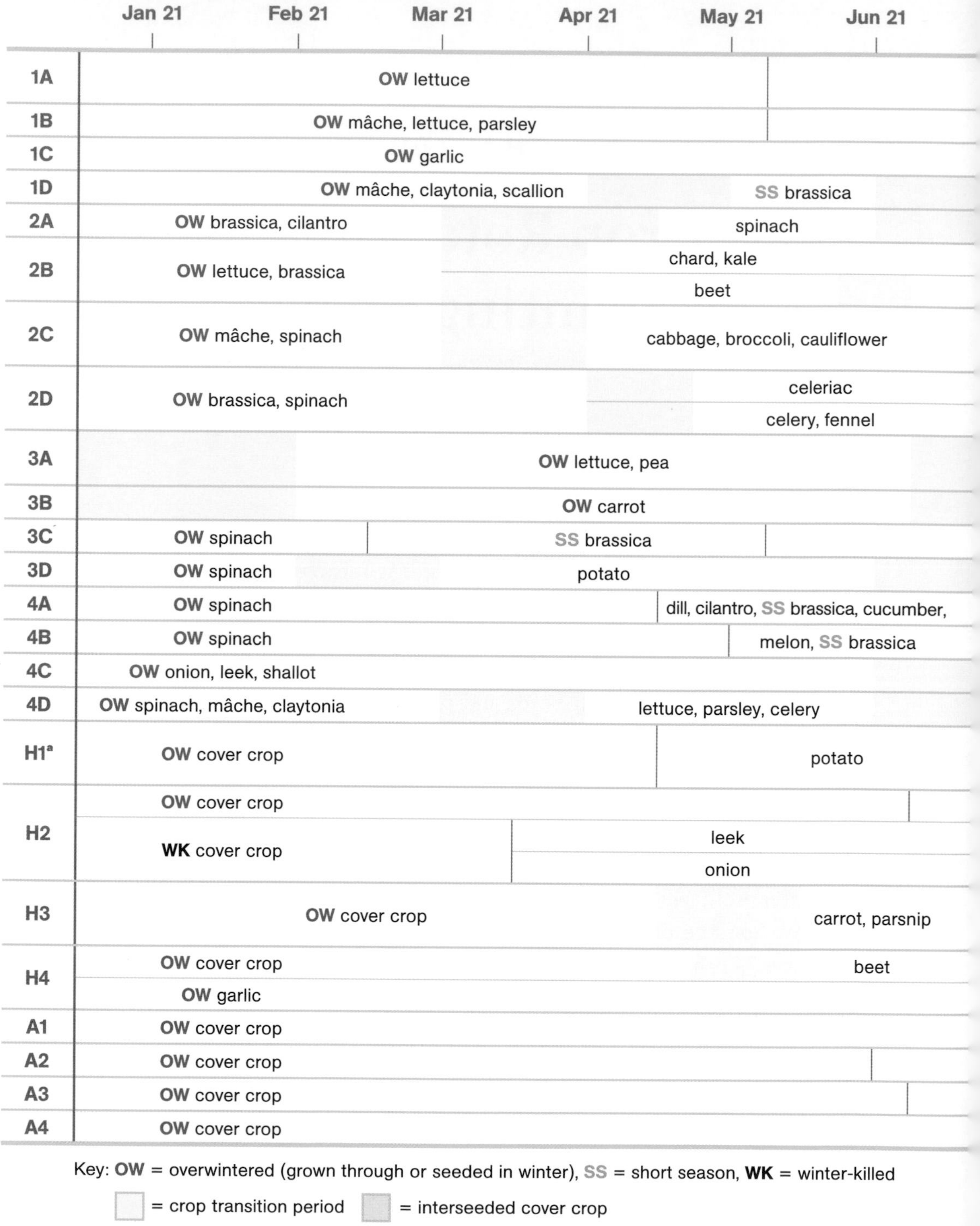

	Jan 21	Feb 21	Mar 21	Apr 21	May 21	Jun 21
1A	OW lettuce					
1B	OW mâche, lettuce, parsley					
1C	OW garlic					
1D	OW mâche, claytonia, scallion				SS brassica	
2A	OW brassica, cilantro				spinach	
2B	OW lettuce, brassica				chard, kale / beet	
2C	OW mâche, spinach				cabbage, broccoli, cauliflower	
2D	OW brassica, spinach				celeriac / celery, fennel	
3A	OW lettuce, pea					
3B	OW carrot					
3C	OW spinach		SS brassica			
3D	OW spinach			potato		
4A	OW spinach				dill, cilantro, SS brassica, cucumber,	
4B	OW spinach				melon, SS brassica	
4C	OW onion, leek, shallot					
4D	OW spinach, mâche, claytonia				lettuce, parsley, celery	
H1[a]	OW cover crop				potato	
H2	OW cover crop					
	WK cover crop				leek / onion	
H3	OW cover crop					carrot, parsnip
H4	OW cover crop					beet
	OW garlic					
A1	OW cover crop					
A2	OW cover crop					
A3	OW cover crop					
A4	OW cover crop					

Key: **OW** = overwintered (grown through or seeded in winter), SS = short season, **WK** = winter-killed

□ = crop transition period   ▨ = interseeded cover crop

FIGURE 7.1. Master planting chart 2018–2019.

Crop Rotation and Planting Cycles

Jul 21	Aug 21	Sept 21	Oct 21	Nov 21	Dec 21

- tomato, cilantro, dill | **SS** brassica | **OW** onion and leek seedbeds
- tomato, basil | **OW** spinach
- brassica, fennel, celery, lettuce | **OW** spinach
- lettuce, cilantro, celery, chard | **OW** spinach, mâche, claytonia
- squash, cucumber, bean, cilantro, dill | **OW** garlic
- **OW** lettuce
- basil
- **OW** parsley
- chard, basil, cilantro | **OW** mâche, lettuce
- **OW** mâche, claytonia, scallion
- lettuce, brassica
- cucumber, squash, sweet corn, cilantro, arugula | **OW** mâche, spinach
- lettuce | **OW** brassica, cilantro
- pepper, eggplant, okra | **OW** lettuce, brassica
- lettuce | **OW** spinach, brassica
- squash | lettuce | **OW** spinach
- brassica, cilantro, spinach | **OW** lettuce, pea
- lettuce | pea greens | **OW** carrot
- cilantro, lettuce | **OW** spinach
- **OW** garlic
- cabbage, cauliflower, broccoli
- brassica
- **WK** cover crop
- turnip, radish, bok choy
- brassica
- winter squash
- melon, tomato
- carrot, beet

Note: Sections 1A–4D are approximately 3,000 square feet (280 sq m) each. Sections H1–H4 and A1–A4 are approximately 10,000 square feet (930 sq m) each.

The changeover from one crop to the next may take place immediately, such as mowing the beds of overwintered spinach and then sowing short-season brassicas (3C). Or the transition may take two weeks or longer as we gradually harvest and replant. For example, the transition from overwintered carrots to summer-planted lettuce (3B).

[a] We interseed cover crops beginning at the end of August with crimson clover, followed by a second seeding in September of winter rye, vetch, and other cover crops. In half of the carrot and parsnip planting (H3), we interseed a crop that will winter-kill to allow for early-spring planting of leeks and onions the following year.

What to Plant and Where

In some cases the chart calls for several crops to be planted in one growing area during a particular part of the growing season. This gives us an overall framework with some built-in flexibility in space allocation, which allows us to make adjustments in response to market demands. For example, early cucumbers are planted in one growing area along with summer squash, cilantro, and some brassicas. If there aren't enough cucumbers to meet the early demand one year, we increase the amount of space in the growing area for the first planting of cucumbers the next year by cutting back on summer squash or adding cucumber to another area.

We plant multiple vegetable successions in a growing area in a given season. A bed of overwintered lettuce, peas, and beets will be harvested during the spring, for example, and then basil, beans, more lettuce, and late cucumbers and squash will fill the space. After those crops are harvested, mâche and spinach will be planted and covered with a low tunnel for harvest in the winter and following spring. Many variables determine when the harvest of a particular crop will be finished, and thus our plan is flexible as to when the next crops will be started. Some crops, however, need to be started at a specific time in order to achieve consistency of production through the season, and the plan can be altered according to those needs. In other words if a certain crop has to be planted in early July, but the planned-for bed space is not yet available, we just plant the crop in any appropriate bed where there is an opening. Since the fields are full of such a wide range of vegetable crops and cover crops, there is bound to be a crop somewhere that can be harvested or mowed down so time-sensitive crops can be put in. Not only does this give us the flexibility we need, it also keeps the fields in consistent vegetable cover, which is beneficial for soil health and profit.

We also use the chart to figure out where transplants will be required and how many. This gives us our transplant production schedule. Transplants are generally overproduced, which ensures that we always have enough to fill the allocated space as well as the freedom to set out only the best-looking plants. We can set out more plants than we originally planned to if a reason presents itself.

Here is a simple example of how to figure transplant production: The areas on the chart allocated to tomatoes equal six 120-foot (37 m) long beds. At a 2-foot (60 cm) spacing for each tomato, it takes 60 tomatoes to plant 120 feet.

60 plants per bed × 6 beds = 360 tomato transplants

In this case, we would grow at least 400 transplants to full size, select the best for planting, and give away the rest.

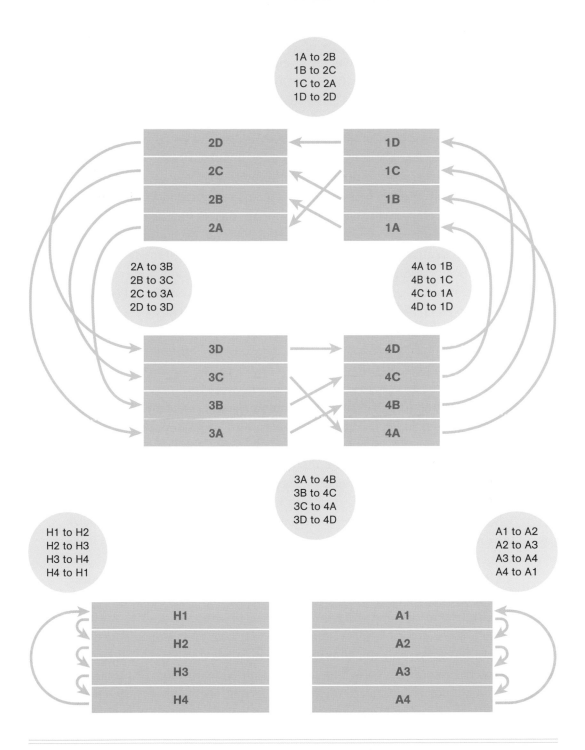

FIGURE 7.2. Crop rotation map. In Field H and Field A, the rotation is straightforward. Each year the crop group moves to the next bed: A1 to A2 to A3 to A4, and then back to A1. The pattern in Field H is the same. In our home field, where we grow crops year-round using low tunnels for winter protection, the rotation pattern is more complex, as shown here.

Selecting which crops to grow is based on many factors, such as market demand, how the crops grow in the given environment, when and where they can be grown, and grower preferences. Diversity is generally best in this regard, both for the soil and to test whether there have been changes in the growing environment, the grower's production system, or market dynamics that will improve the profitability of the various crops. Diverse crop rotations along with interseeding or interplanting crops help to maintain flexibility in the soil life, which will further support the successive vegetable plantings.

How well the crop grows is a primary factor in its selection for cultivation. The ability of a crop to produce well is determined by factors such as soil conditions and environment, time of year, and skill of the grower. Each farm has growing conditions that favor certain crops, and growers will do well to identify those crops. Conditions can vary even within a single field, and thus some crops may do better than others. At our farm, the D sections of the home field (the year-round production field) are in wetter, shadier conditions. Thus we select crops suited to those growing conditions for these areas, and rotate the crops through the areas accordingly. Summer lettuce in these areas is a prime example.

There are other ways to match crops to conditions, too. For example, summer lettuce might be selected to grow in the shade between rows of tomatoes. Along these lines comes the practice of companion planting, or planting crops that benefit one another, such as the Native American Three Sisters planting of field corn, winter squash, and pole beans. Many factors determine how crops will benefit or be of detriment to one another, so growers need to pay attention to their specific conditions. Our fields are generally nitrogen-rich and watery with much growth force. In this case a leguminous cover crop fixing more atmospheric nitrogen may well be detrimental to the next crop instead of beneficial. Instead, a coarse carbon- and silica-rich grass cover crop may be best to precede a vegetable crop.

Crops are historically rotated by plant families in order to avoid weed, insect, and disease pressures. However, when these conditions are dealt with through soil fertility and balance, there is much less need to set up rotations on this basis. Actually if a crop grew very well in a particular bed, most likely immediately replanting or reseeding the same crop in that same soil will yield another very productive result. This follows nature's example of not rotating plants quickly. If the soil biology is in place to grow a crop well, the plants therein created conditions of symbiosis with the soil life. This soil life is then well prepared to carry on a second crop. So in a nutshell, if a crop grew well, there is no need to rotate. If there were pest issues, rotation may be helpful.

In terms of classical crop rotation to reduce pest and disease pressure, there are nine botanical families of crops that make up most of the popular crops grown on vegetable farms and gardens.

Brassicaceae: Broccoli, kale, bok choy, brussels sprouts, cabbage, turnip, rutabaga, mustard, arugula, radish, and so on

Amaranthaceae: Beet, chard, spinach (formerly Chenopodiaceae)

Amaryllidaceae: Onion, shallot, garlic, leek, scallion (commonly referred to as the alliums)

Cucurbitaceae: Melon, squash, cucumber

Poaceae: Corn and grains, grasses (also called Gramineae)

Fabaceae: Beans, peas, clover (also called Leguminosae)

Solanaceae: Tomato, pepper, eggplant, potato

Apiaceae: Carrot, parsnip, parsley, celery, dill, cilantro, chervil (also called Umbelliferae)

Asteraceae: Lettuce, endive, chicory, dandelion, burdock, salsify (also called Compositae)

A few crops—okra, basil, mâche, and claytonia—belong to still other families. Many diseases and insects impact only a specific plant family, so rotation by plant family may be useful if pest pressures are impacting crop productivity. Incorporating plants from the different plant families in a planting plan is useful in terms of encouraging overall diversity of life.

Determining When to Plant

The time when a crop is grown also has dramatic influence on its productivity. As discussed in earlier chapters, sunlight availability and temperature among other environmental factors have a dramatic impact on crop growth. If a crop is very heat loving like eggplant, and the plants are set out too early they will not recover and yield to the extent that a later planting will. This is true of many heat-loving crops. The same is true of cool-loving crops. Spinach seed will germinate in our soils in August, yet these early seedings do not have the vigor and yield of later seedings in September and October. So most crops have particular times of the year best suited to their growth. However, growing these crops slightly out of their season is often the most profitable avenue, due to high demand. All the more reason to have environmental and soil conditions in place that can support such off-season growth.

The amount of sunlight and warmth is the primary factor that determines the best time of year for a particular crop to grow; there is also the moon's dramatic influence on growth to consider. In terms of seeding crops, we pay attention to three moon dynamics: the proximity, the declination, and the phase of the moon. The proximity of the moon to the earth is determined by its elliptical orbit and has a cycle of 27.5 days. When the moon is moving closer to the earth is potentially a time of benefit for seeding. The moon's influence on crops and many other conditions like tides is greatest when it is in closest proximity (at perigee) to the earth. The declination is the height or angle of the moon as it moves across the sky. The moon's pattern of movement is very similar to the sun's apparent movement in this regard, but instead of taking a year to complete a cycle, the moon completes its cycle in 27.3 days. When the moon is ascending in the sky, similar to the sun in spring, it is generally regarded as a period for seeding.

The phases of the moon are probably the most traditionally watched indicator for the seeding of a crop. The period from the new moon to full moon (waxing period) is considered best for the start of growth. If all three moon factors line up positively—waxing, ascending, and approaching perigee—then it may be best to reduce seed rate. If the moon is waning, descending, and approaching apogee (farthest distance from the earth), then it may be time to increase the seed rate. To some degree, we try to carry out seeding during more favorable lunar conditions, but sometimes a crop will need to be seeded regardless of lunar position. There are also the celestial positions of the sun and moon as well as the other planetary bodies that are influencing conditions; here a study and understanding of astrology can be useful.

CHAPTER 8

Soil Fertility and Crop Health

*S*oil fertility refers to the overall ability of a soil to produce abundant healthful crops. As growers it is one of our primary duties to manage the soil to bring it into an increasing state of fertility. Factors that work against this effort include pollution, disturbed weather conditions, insufficient understanding of agricultural practices, and the lack of farm profitability, which can lead to an inability to invest in soil-enhancing methods and materials. Yet there is also much in our favor because the earth actively seeks to improve the soil's fertility. The aware grower can assist this flourishing of life to an astonishing extent. The degree of a soil's fertility is related to the function of the biological, chemical, physical, and energetic states of the soil, which growers can observe and, to some degree, measure. The observation and testing of soils and plants gives the indications for appropriate materials that can be applied by the grower in order to assist soil fertility.

Organic Matter and Fertility

The level of organic material in the soil has a dramatic impact on soil fertility, as well as the grower's soil practices. The consistent physical examination of soil for signs of organic matter accumulation gives the grower the ability to observe how management practices are impacting soil life. As soil organic matter levels increase, many benefits will potentially result if the grower understands how to respond. Increased organic matter content usually signals an increase in soil life, which in turn will result in faster decomposition of materials and greater nutrient release from both the organic and the mineral portions of the soil. In addition, there will be greater fixation of atmospheric nitrogen into molecular forms embedded in the soil life. This soil life thrives on large volumes of food in the form of plant residues, mulches, composts, and plant root exudates, and it's a primary duty of a grower to be an ample provider! This food for the soil life is often the heaviest and most voluminous fertilizer material growers apply to their fields. It is best to keep the soil life evenly fed in order to achieve steady nutrient release and growth of crops. To ensure this steady nutrient release, attention to the efficiency and rate of surface residue decomposition is essential. Abundant diverse materials in an active state of decomposition are a beneficial condition. The increase of organic matter also dramatically impacts the soil porosity, with all the impacts of this condition on soil, air, water levels, and soil life diversity. Again this can be of great benefit to a grower, but very high porosity may require different management techniques. Often it is said an organic matter content of 5 percent or more is beneficial to vegetable production; many vegetable fields and gardens may have significantly higher levels. The ramifications of the grower's management of soil life were fully covered in earlier chapters.

Crop and Soil Assessment Tools

In addition to careful observation, there are tools that provide quantifiable assessment of growing conditions. By giving measurable results, they help train growers in the finer details of soil and crop health observation. Growers can purchase devices such as penetrometers, Brix meters, pH meters, and electrical conductivity meters to perform in-field assessment of conditions. This is an inexpensive way for growers to obtain immediate information on a wide variety of growing areas and crops. A more thorough assessment can be achieved through laboratory analysis of soil and crop tissues. The combination of in-field testing, laboratory analysis, and particularly careful physical examination of growing conditions provides interrelating guidance for grower actions regarding soil fertility efforts.

IN-FIELD CROP AND SOIL ASSESSMENT TOOLS

A penetrometer is used to assess soil compaction. It is essentially a steel rod attached to a pressure meter. This tool allows a grower to quickly assess the compaction of a soil at various depths. The pressure required varies as the rod is pushed through the soil layers. These pressure changes indicate potential compaction pans. The information gained is not as complete as an assessment carried out by digging through the soil to directly observe the relative compaction of the layers, but it is certainly much quicker to push down a rod than to dig a hole. A homemade version is a steel rod with a bent handle at one end, essentially forming an L-shape. With this tool growers use their own senses to

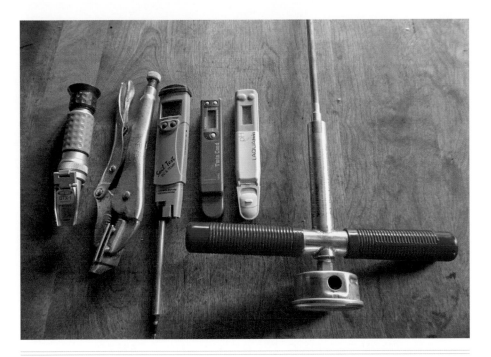

FIGURE 8.1. On-farm plant and soil assessment tools include a refractometer and adjusted vise grip for extracting and assessing plant sap, three types of electrical conductivity and pH meters for both soil and sap analysis, and a penetrometer for testing soil compaction.

determine when the pressure needed to push the rod through the soil layers signals a compaction issue.

It is relatively easy to test soil pH using an inexpensive pH meter with a direct-reading soil probe. Another simple method is to mix distilled water with a soil sample and use pH test strips to check the pH. There are also meters that measure the pH of liquids. These meters are capable of reading the pH of not only soil / distilled water mixtures but also extracted plant saps. A purchased modified vise grip plier is a helpful tool to extract plant saps for testing.

Meters are also available that measure electrical conductivity (EC) of soil, which provides an indication of the volume of simple ions that may be available for uptake by plants. This not only gives an idea of the potential need for fertilization, but is also helpful in determining whether the volume of fertilization materials applied was insufficient, appropriate, or excessive. Like pH meters, EC meters are available with a direct-reading soil probe, or there is a model that gives a measurement of liquids. The liquid-measuring meter requires mixing soil with distilled water and is also capable of measuring plant saps.

In terms of using plant data to assess soil fertility, measurements of Brix levels, plant sap pH, and plant sap EC can be revealing. Brix is a measurement of dissolved solids in plant sap, and it can be a helpful indicator of plant vitality. All these meters can be used right in the field to take measurements. Many variables affect Brix readings, but in general the higher the reading the sweeter the flavor of the vegetable. Higher readings usually correlate with many improved plant qualities as well. Along with in-field observation and soil testing, the Brix, EC, and pH of plant sap can provide further information about fertilizer excesses and deficiencies as well as guidance toward which kinds of fertilizer may be of most assistance. As with most testing procedures, the more plant sap testing is done over time and in various conditions, the better. Frequent data collection allows growers to identify trends, which are more meaningful than an individual numerical reading. For instance, a grower can test the EC and pH of plant sap and soil as well as the Brix of the sap, then add fertilizer material to the soil and, after allowing an appropriate time for fertilizer integration, retest. The changes in these parameters may yield insights into the effectiveness of that fertilizer material, and when such tests are repeated under varying conditions with similar results, growers can adjust fertilization efforts accordingly.

LABORATORY TESTING OF SOILS AND PLANTS

Laboratory testing of soils and crops is quite diverse. There is soil testing for chemical constituents, biological factors, and textural characteristics as well as crop testing of saps and various tissues. All of these tests as well as on-farm testing with various tools offer the potential for insights, but again what is often of more importance is the direct observation of soil and crop conditions. The lab test results or in-field meter readings are for general insight and further development of the grower's ability to see what a crop requires, but are not meant to override what may be indicated from field conditions. As growers develop their abilities to physically assess crops and soils, these testing procedures become dispensable to some degree. Lab tests of soil samples frequently are utilized to assess the mineral materials of the soil, whereas laboratory analysis of plant tissues and saps often offers a measurement of nutritive elements that plants have been able to take up from the soil, which can differ from the constituents that make up the mineral materials of the soil. Some universities, as well as some private labs, offer leaf tissue analysis of elemental nutrients, and it is often quite affordable. Sap analysis is a more recent development and is offered mostly by private laboratories. Performing some form of sap or tissue analysis in conjunction with laboratory soil testing can provide valuable data.

The analysis of soil samples by laboratory testing methods is an attempt to quantify various measurable characteristics of the soil. Though this is an exciting field of human development and is sure to continue to progress in usefulness, it has historically been viewed as of somewhat limited use by growers. This is likely due to the difficulty of accurately assessing relatively large areas of soil, often an acre or more, from the very small sample of soil sent to a lab. In addition, soil characteristics such as levels of extractible nutrients can vary depending on time of year, weather conditions, soil temperatures, and other variables. However, soil testing can be useful for tracking mineral or biological trends in a general or approximate manner. In this regard the more testing is done over a longer period of time, the greater the usefulness. Consistently using the same laboratory to run the analysis and taking soil or tissue samples from the same area and at the same time of year can provide better data for making meaningful comparisons and seeing trends developing over time. This focus on long-term assessment keeps growers from overreacting to an occasional erratic reading on a single year's test.

The most common type of soil analysis is the use of various extraction chemicals and procedures to assess the nutritive elements contained in the sample. The variance in laboratory materials and techniques is why test result data do not stand up to cross comparison between different soil testing labs, and this is another reason why it is best to consistently work with the same laboratory. The choice of extraction agent used in the soil test is important. Some tests use a strong acid to extract nutrients; other use a weak acid or water.

Soil testing with these various agents extracts only a percentage of the total of a given element from a sample. Because of this, testing a soil sample with both a strong acid and a weak extracting agent can offer additional insights. Some labs offer this, or samples can be split and sent to separate labs for results. The strong acid extraction can show which nutrients would be available if there is strong biological function in the soil, whereas a weak extraction will show what nutrients are highly available to the plant, even if soil biology is weak. Thus the results of the two tests often do not correlate to each other. With many soils having low levels of biological activity, the results of the weak extractions are often much more aligned to the results of plant tissue analysis of crops grown on that soil area. The strong acid tests show which nutrients can or cannot become available with increasing biological activity. If an individual element is deficient in both strong and weak acid tests, a broad-scale fertilizer application of significant volume to supply that nutrient would likely be in order. However, if the nutrient is deficient in the weak acid extraction but sufficient in the strong acid, then perhaps a small-volume application of an appropriate highly plant-available fertilizer, combined with efforts to increase biological activity, may be the best approach.

The soil lab compares the extracted volume of nutrient with other samples that have proven sufficient for crop production under various conditions in the past, and recommendations are based on this set of information. Consider recommendations from the soil laboratory as a helpful guide that is perhaps not fully authoritative. They are particularly useful as a source of ideas for newer

CROP SERVICES INTERNATIONAL, INC.
29246 Lake Street, Marcellus, MI 49067
www.cropservicesintl.com

Field ID: 2
Crop: Vegetables
Acre: 1 acre

Test Date 12/10/18	Base Saturation		CEC (Mehlich 3) in lbs/acre		Saturated Paste Test in ppm	
	TARGETS	RESULTS	TARGETS	RESULTS	TARGETS	RESULTS
pH	–	6.6	–	–	6.0–6.3	6.6
CEC	–	11.9	–	–	–	–
Organic Matter %	–	9.9	–	–	–	–
Chlorine	–	–	–	–	< 60	7
Salts	–	–	–	–	< 1200	72
Bicarbonate	–	–	–	–	< 90	70
Sulfur	–	–	80	42	5–6	2.1
Phosphate	–	–	572	750	0.3–0.4	0.41
Calcium	–	–	3225	3452	20–40	10.4
Magnesium	–	–	341	349	4–8	2.8
Potassium	–	–	370	316	10–12	12.0
Sodium	–	–	109	39	< 5	1.5
Calcium %	68	72.8	–	–	60	46.1
Magnesium %	12	12.3	–	–	20	20.6
Potassium %	4	3.42	–	–	12	27.5
Sodium %	< 3	0.72	–	–	< 5	5.9
Other %	4	4.8	–	–	–	–
Hydrogen %	10	6.0	–	–	–	–
Boron	–	–	4	1.16	0.05–0.1	0.1
Iron	–	–	150	288	0.5–1.5	0.6
Manganese	–	–	100	36	0.07–0.1	0.03
Copper	–	–	10	8.56	0.05–0.08	0.02
Zinc	–	–	37	30.06	0.07–0.15	0.02
Aluminum	–	–	–	2648	< 1.5	0.83
Optional Tests	Cobalt	Molybdenum	Nickel	Selenium	Silicon	Carbon
TARGETS	1–2 ppm	1–2 ppm	1–2 ppm	1–2 ppm	35–45 ppm	5–12%
RESULTS	0.064	0.43	–	–	–	–

FIGURE 8.2. Soil test results.

growers who are not aware of the diversity of options available to them. Many private soil labs have tested a large number of soils over the years and may be more aware of the full range of fertilizer options than individual growers. This level of experience can be enlightening, though caution may be needed if the soil lab is also a fertilizer dealer or the result may be unnecessary expenditures. Soil labs based in academic settings do not sell products, but are often unaware of all the fertilizer options available in the private marketplace. There are also privately run labs that are knowledgeable about fertilizer materials but do not sell them. It is worth spending the time researching the methods that labs use and speaking with other growers about which labs they choose to work with. In the end, though, the choice of the appropriate lab is up to the individual

UCONN SOIL NUTRIENT ANALYSIS LABORATORY
6 Sherman Place, Unit 5102, Union Cottage
Storrs, CT 06269-5102
www.soiltest.uconn.edu

PLANT ANALYSIS RESULTS

Client: Tobacco Road Farm	Date Received: 9-11-2018	Date Processed: 9-19-2018
Crop: Carrots	Sample ID: Carrot field #2	Lab ID: T18-193

Plant Nutrient	Sample Results	Sufficiency Range
Nitrogen (N) % Dry Weight	3.15	2.1–3.5
Phosphorus (P) % Dry Weight	0.65	0.2–0.5
Potassium (K) % Dry Weight	4.3	2.5–4.3
Calcium (Ca) % Dry Weight	0.94	1.4–3.0
Magnesium (Mg) % Dry Weight	0.27	0.30–3.0
Boron (B) PPM Dry Weight	32.80	30–100
Copper (Cu) PPM Dry Weight	10.30	5–15
Iron (Fe) PPM Dry Weight	65.00	50–350
Manganese (Mn) PPM Dry Weight	42.40	60–300
Molybdenum (Mo) PPM Dry Weight	29.40	No data
Zinc (Zn) PPM Dry Weight	43.80	25–250
Non-Essential Elements		
Sodium (Na) % Dry Weight	0.23	No data
Aluminum (Al) PPM Dry Weight	24.50	No data
Lead (Pb) PPM Dry Weight	0.60	No data, Ideal value would be 0

Values based on sample consisting of 20 mature leaves from new growth. Summer.

From: Jones, J.B. *Plant Analysis Handbook*. Athens, GA: Micro-Macro, 1991.

FIGURE 8.3. Tissue test results.

grower. Each grower's individual situation is unique. The time invested in this is well spent, because the relationship with the lab will hopefully be a long-lasting one that provides an extended collection of relevant information about the grower's soil.

To better describe how analysis data can be useful, let's look at some actual soil test results. Figure 8.2 shows soil test results from one of the fields at Tobacco Road Farm and provides an example of the use of a strong acid Mehlich-3 extract and a saturated paste (water) extract of the same soil sample. Many soil test reports list the organic matter content of a sample as a percentage of total soil volume. Soil samples are usually screened before testing to remove large particles of organic matter, so the reported percentage of organic matter is a reflection of smaller fragments of undecayed materials, small living organisms, and decomposed materials (some of which are humic substances). The level of organic matter dramatically influences the forms of the measured nutrient elements of the soil test. This is often of relation to their availability for plant growth, as reflected by the saturated paste test. The organic matter content, especially in the form of humic compounds, also impacts the nutrient-holding capacity of the soil, which—along with the clay materials in the soil—forms the basis for the calculation of the cation exchange capacity as stated on the soil test.

ABOUT CATION EXCHANGE CAPACITY (CEC)

The measurement designated as cation exchange capacity (CEC) is an attempt to quantify the nutrient-holding capacity of a soil. Although it is expressed as cation (a positively charged ion) holding capacity, this soil characteristic also is related to the soil's ability to hold onto anions (negatively charged ions). Table 8.1 summarizes the common anions and cations in soil.

CEC is a measurement of the chemical reactivity of humic and clay materials, and thus its level is highly related to the organic matter content and texture of the soil. Basically, the higher the CEC, the greater the capacity for the soil to hold nutrients for plant and microbe uptake. In general a higher CEC is desirable because it indicates that the soil offers more volume of available nutrient to the living system. Sandy soils often have a relatively low CEC because of low clay content and the accompanying low organic matter content. Yet with the addition of organic materials and perhaps clay, sandy soils improve greatly and are generally viewed as superior for the culture of many vegetables. Clay soils often have relatively high CEC levels, however they also benefit from organic matter additions in order to increase porosity, which creates more aerobic conditions and thus improved biological activity.

Raising a soil's CEC or nutrient-holding potential can be difficult on extensive vegetable acreage, but it is generally achievable in intensive vegetable production systems. The smaller land area in cultivation combined with the very high value of the crop allows for much greater input potential than in other farming systems. Application of 30 tons or more of compost per acre (74 metric tons per ha) combined with clay mineral additions, reduced tillage, and careful soil management can raise CEC numbers relatively quickly. One minor difficulty of high CEC soils is that they can present growers with some difficulty in adjusting elemental nutrient balance. Consider the example of a high CEC soil in which the soil colloids hold a large volume of a particular element. Bringing this large volume into balance with other elements will require application of much larger volumes of amending materials than if the soil had a low CEC.

The estimated balance of the various cations of the soil is referred to as base saturation percentage. The base saturation is a calculation of the pounds per

TABLE 8.1. Common Cations and Anions in Soil

Cations	
NAME	**FORMULA**
Hydrogen	H^+
Potassium	K^+
Calcium	Ca^{2+}
Magnesium	Mg^{2+}
Sodium	Na^+
Zinc	Zn^{2+}
Copper	Cu^+, Cu^{2+}
Manganese	Mn^{2+}
Iron	Fe^{2+}, Fe^{3+}
Cobalt	Co^{2+}
Anions	
NAME	**FORMULA**
Hydroxide	OH^-
Phosphate	$(H_2PO_4)^-$ and other forms
Sulfate	$(SO_4)^{2-}$
Nitrate	$(NO_3)^-$
Borate	$(BO_3)^-$
Molybdate	$(MoO_4)^-$
Chlorine	Cl^-

acre of a nutrient extracted combined with the CEC measurement to arrive at ratios of nutrients to one another. Soil test results will report the base saturation percentage of calcium, magnesium, potassium, sodium, hydrogen, and other bases. General target percentages are often set at 60 to 70 percent calcium, 10 to 20 percent magnesium, 3 to 5 percent potassium, less than 3 percent sodium, 10 to 15 percent hydrogen, and 2 to 4 percent other bases. Comparing cation levels in this way may result in some useful information, but the objective is not to balance a soil test but to get results in the field; these two goals often align, but not always. Strong biological activity can overcome chemical imbalances that are measured by a CEC test. Physical characteristics of a soil may also alter these target percentages. For example, a higher Ca percentage and lower Mg percentage (than target values) may be better for a clay soil because of the loosening of soil structure aided by calcium. Vice versa; a higher magnesium percentage and low calcium percentage is favored for high organic matter soils where soil structure could use the tightening offered by magnesium. A measurement we find particularly useful is potassium levels; deficient potassium levels can be as detrimental as excesses in terms of healthy crop growth. Sodium (Na) is often included in base saturation percentages but not always. Generally sodium in excess is more problematic than sufficiency, though sodium is very influential in the flavor of vegetables and is therefore worth paying attention to. The hydrogen percent is a reflection of the acidity (pH) of the soil. A 10 to 15 percent hydrogen base saturation equates to a pH of about 6 to 6.5, which is a slightly acidic soil favored for vegetable growing. Two to 4 percent other bases refers to the unaccounted-for multitude of other bases in the soil; usually it helps everything add up to 100 percent.

INTERPRETING SOIL TEST RESULTS

The majority of the data on a soil test report deal with the measurements of individual nutritive elements. Many labs provide a target number, which they calculate by various means, that is deemed appropriate for the given soil. These separate measurements of essential nutritive elements do provide some clues as to how to approach fertilization, and this information is potentially useful to growers, but it is not necessarily accurate given the variables of a living soil. The goal is not to fertilize to reach arbitrary target quantities of mineral elements, but to produce high-quality vegetables in the field. What a soil test cannot fully capture is the ramifications of the environment in the soil, in which the elemental nutrients are parts of larger compounds in living systems of great interconnection and complexity. This is a critical consideration because the addition of one nutrient may lower or increase availability of another due to the impact on soil chemical, physical, biological, and energetic conditions. The addition of some mineral fertilizers may even lower the availability of the elements of which they are made! This is often due to the damage to the soil biology that a fertilizer material can potentially cause. However, the separation and measurement of the essential nutritive elements on a soil test does provide some clues as to how to approach fertilization.

Managing the Nutrients

Although mineral nutrients are interconnected and influence one another's availability, it is useful to examine each one to better understand individual deficiencies and excesses. This understanding can then help lead to better conditions of balance for the whole as grower actions are modified in response to these observed and measured conditions.

Hydrogen or pH is often measured on a laboratory soil test but can also be easily measured using on-farm soil analysis with various meters, or pH test strips and distilled water mixtures. pH attempts to quantify the measurement of the ratio of hydrogen ions (H+) to hydroxide (OH–) ions. Understanding the pH of a soil can be useful in determining potential elemental nutrient availability. The hydrogen ion's dominance aids acid formation, which can help reactions that solubilize needed nutritive elements. Many cations are generally more available in somewhat acidic conditions. The makeup and functioning of soil life are also indicated by general soil pH reading. Higher pH readings tend to favor a higher level of bacteria and thus a faster-moving nutrient release from decomposition. However, a more acidic soil may have a higher level of fungal activity, giving a release of much-needed nutritive elements held in bonds that would be insoluble to bacterial action. For many vegetables a pH reading of about 6.5 is considered optimum. Thus the pH gives a reading of the general environmental condition of a soil, both biological and chemical. The pH is where these two worlds meet on a soil test, and it is important to understand this interrelationship, as either can create conditions of acidity or alkalinity. It is also important to remember that the soil pH is a general measurement of a blended soil and that the biology of the soil rhizosphere or areas of specific biological activity can be very different from the overall soil reading. This is a demonstration that specific biological conditions are developed by nature for specific soil conditions to release specific nutrients for the benefit of life.

NITROGEN

Soil nitrogen (N) is very difficult to assess by soil testing due to its transient nature. This is unfortunate because nitrogen management is very important for crop growth. The most common forms in the soil are nitrate (NO_3), ammonium (NH_4), and nitrogen embedded in organic forms such as amino acids and proteins. When there is too little N in the soil, crop yields are generally low; too much nitrogen and crops are too lush and prone to insect and disease attack.

The signs of N deficiency are slow, restricted growth; yellowing of leaves, especially younger ones; and withering of older leaves (leaves may drop off). Growers can employ tissue analysis to measure the nitrogen content of plants, which often relates to available N in the soil. Nitrogen is primarily supplied to living systems through the fixation of atmospheric nitrogen by soil microbiology. Biological nitrogen fixation can be substantial in highly functioning soils and is the forte of the legumes, with their symbiotic relationship with rhizobium bacteria. In addition to this natural fixation of nitrogen, a vast amount of artificial nitrogen fixation takes place in fertilizer factories. In this process, nitrogen gas is combined with hydrogen from methane (natural gas) to form ammonia (NH_3). The ammonia is then converted into various synthetic nitrogen fertilizers as well as industrial chemicals. This nitrogen is produced in staggering quantities and is the cause of much pollution due to its extensive application to the earth's surfaces. This is the source of most of the soluble nitrogen fertilizer utilized in agriculture. These soluble inorganic forms of nitrogen are generally very disturbing to the delicate balance of a soil, but it is up to growers to determine if this is indeed the case for the particular soils they manage. The addition of nitrogen-rich materials to soils deficient in nitrogen will often greatly increase biological activity, which results in an associated release of many various nutrients for plant uptake.

Excessive nitrogen levels are a relatively common occurrence on vegetable farms, because many growers view nitrogen fertilizer as the primary means to produce a quick, large, weighty crop. The excess of nitrogen can cause deficiencies in other critical elemental nutrients such as magnesium, silica, and copper as well as cause imbalances in protein formation within the plant. Excesses can also lead to elevated levels of nitrate in the plants, which can cause all manner of health difficulties for those who consume such plants. Nitrogen compounds are common nervous-system exciters, and excess is a common cause of agitation, anxiety, heart irregularities, and a host of other difficulties in humans. Because of the abundant production and application of inexpensive synthetic nitrogen, ailments due to excessive nitrogen are very common. Fortunately nature sends in the insects and disease to destroy such crops before they are consumed by humans, unless of course growers apply pesticides to defend and preserve them . . .

Nitrogen is not properly processed within living systems unless sufficient amounts of other critical nutritive elements are also available to plants. This is one of the most important functions of a vibrant, high-functioning soil life: the ability to supply necessary nutrients in balance so that the levels of nutrients in plant tissue are appropriate for vibrant growth. Some particularly important elements in nitrogen balancing are sulfur, copper, cobalt, manganese, and molybdenum as well as carbon. Sulfur is a critical component of several amino acids along with nitrogen. Copper, cobalt, manganese, and molybdenum are all critical components in the reactions that put nitrogen into forms that can be incorporated into the organic living realm. Shortages of these nutrients in plants can cause a buildup of inorganic forms in the plant matter, with its associated problems of insects and disease.

Though balance of all nutritive elements is critical and is likely why the organic method of fertilization has been successful in the past, the most important specific elemental balance for a vegetable grower to monitor and manage is probably that of carbon to nitrogen. Managing a balanced flow of carbon- and nitrogen-rich materials into the soil system is an essential technique for vegetable growers. C:N ratios are often discussed in terms of compost pile assembly, but the grower will greatly benefit from also carefully considering the carbon to nitrogen ratio of all the inputs that are entering the soil. This includes fertilizers, composts, mulches, crop residues, root exudates, cover crops, water sources, and more. Nitrogen inputs are generally of the "green" nature, often sloppy and smelly, whereas carbon inputs are often brown, dry, and hard. Common carbon inputs are dried leaves, straw, wood chips, ligninrich crop residues, sugars, charcoals, and humates. Another source is carbonrich sugar created through photosynthesis and exuded into the soil system by plant roots. Carbon inputs can go a long way toward reining in excesses of nitrogen in the soil because they are a source of materials that soil biology can bind with the nitrogen. Applying too much carbon may of course induce a nitrogen deficiency in the system, with the result of a decreasing crop yield. Maintaining a balanced flow of carbon and nitrogen overall leads to an abundant, balanced, well-fed soil microbiology with the associated appropriate nutrient releases for healthful crop growth.

PHOSPHORUS

Many enzymes critical to metabolic functions require phosphorus. These functions include energy and nutrient transfer, sugar and starch formation, flowering, and fruit and seed production. Phosphorus (P) is traditionally known for its importance in increasing the number of fruits and seeds as well as hastening their maturation. A deficiency of phosphorus is often seen as a reddish tinge on older leaves. Other signs may be diminished fruiting and seed set as well as stunted growth in general.

Several variables impact phosphorus availability for crop growth, and thus it is particularly difficult for soil testing labs to assess what constitutes optimal levels of phosphorus across the varied conditions of vegetable croplands. A confounding factor is that labs are often pressured to recommend phosphorus application rates based on the dictates of water-quality regulators rather than on what may be best for crop growth! Tremendous volumes of pollutants are entering our waterways at this time, but government regulators are disproportionately focused on phosphorus contamination. In the face of massive pollution issues due to synthesized nitrogen usage, not to mention all the extremely dangerous chemicals generated as byproducts of industrial production, it would seem that there are much more dangerous materials than phosphorus in need of oversight. Unfortunately regulators currently consider any and all phosphorus-containing materials as potential hazards to water quality, and that includes compost and other organic materials. This is deficient reasoning that fails to take into account how natural systems function. The use of compost and organic materials on land is critical to maintaining them in a state of vitality, which naturally reduces water pollution, including phosphorus. This is not to say that growers should not take care to avoid contaminating water systems. The best way to avoid contamination is by creating vibrant living soils that are not prone to erosion and that support crops without any need for pesticides, and is just what this book is about.

Unlike nitrogen, which is very soluble and can readily leach from soils, phosphorus is prone to extreme stability in the soil; the challenge can be keeping it available for crop uptake. Additions even of soluble phosphate fertilizers are very prone to immediate chemical reactions in the soil that tie them up to a dramatically less available state. This is one of the main reasons why phosphorus is not prone to leaching and the potential for waterway contamination from well-tended soils is so low.

An important key to abundant phosphorus availability is highly functional soil life. The soil life, particularly the fungi, have the ability to solubilize phosphorus from its inert forms. Once the phosphorus is thus brought into living systems, it is more readily cycled from one living organism to another. Much of what is extracted during a laboratory analysis is the phosphorus contained in the living fragment of the soil. Thus, soils high in organic matter usually test higher for phosphorus than soils lacking organic matter, especially when a strong acid test is used. These high levels often indicate a higher level of soil life yet may still not represent adequate phosphorus for crop growth, as that depends on the dynamics of the biology. For a more accurate assessment of how much phosphorus a crop will be able to take up, lab tests using weak acid or saturated paste water extracts may well be more accurate, and often align with tissue analysis of the crop on a given soil.

CALCIUM

Calcium (Ca) is related to cell membrane strength and thus is associated with firmness of fruits and improved storage of vegetables as well as freeze resistance and nutrient movement across cell membranes. Calcium is also known for improving soil structure through a chemical reaction called flocculation, which increases the soil life's ability to aggregate soils. This leads to improved crop root growth in many soils.

Calcium is highly related to crop yield, because supplying adequate levels can dramatically increase the size and weight of a crop. As with phosphorus, strong acid extractions often produce soil test results showing high levels of calcium, but tests using weaker acid extracts and tissue analysis may well show low levels (see figure 8.2 on page 122 for example test results). Again the soil biological function is critical to the steady release of appropriate amounts of calcium through the growing season. Steady release is particularly important with calcium because it is less mobile within the plant than many other elemental nutrients. Deficiencies often show up on the youngest leaf growth because of this poor translocation of calcium. The young leaves are distorted, perhaps twisted or rolled. Growers are well advised to watch for this deficiency, because insufficient calcium is common and can dramatically reduce yield. On the other hand sufficient calcium levels for optimal growth is a sign of high yield potential.

MAGNESIUM

Magnesium (Mg) is the critical element in the chlorophyll molecule, and thus essential for photosynthesis. It is also critical in many other enzymes and enzymatic reactions. Magnesium carries a similar ionic charge as calcium. Magnesium fertilization has less direct impact on crop sizing than N, Ca, and K fertilization. In general the reliance on N, Ca, K fertilization for crop sizing has largely resulted in magnesium deficiencies in both crops and consumers of crops. These three commonly used nutritive elements are all antagonistic to magnesium absorption by crops. The excessive use of nitrogen fertilizers is particularly implicated in magnesium deficiency in humans—to the point of questioning whether humans have magnesium deficiency or nitrogen poisoning. In the highly unbalanced environmental conditions that humans have created, there is much opportunity for growers to exacerbate the imbalances, or to learn how to create conditions for reestablishing balance.

Magnesium's capacity to balance the impacts of N, Ca, K excesses goes a long way toward diminishing the diseases often associated with such excesses. This is a primary reason why sufficient levels of magnesium in weak acid soil extracts and tissue analysis may be an indicator of the reduction of pestilence potential for a crop—it is an indicator of balance. Magnesium's impact on soil structure is the opposite of calcium due to its ability to bring soil particles closer together. This is very useful for sandy and high organic matter soils, which tend to be overaerated. The gluelike effect of magnesium has ramifications for soil aggregate stability and formation. However, excess drawing together of soil particles can also increase soil compaction difficulties from heavy machines or tillage. In this regard the ratio of soil magnesium to soil calcium is important to understand, and observing and monitoring of the impacts of fertilization with these materials on soil structure is important. Magnesium deficiency symptoms are relatively common on vegetable crops, the most common sign being interveinal chlorosis. Lower leaves may yellow brightly and fall. There is also poor fruit development and the potential for overall reduced yield.

POTASSIUM

Potassium (K) is also an important nutritive element for the sizing of a crop, in this case often in the sizing of roots, fruits, and seeds. Some of potassium's roles in metabolism include the activating of enzymes and protein synthesis, as well as the movement of carbohydrate and cell water balance. The relationship with cell water balance makes adequate potassium very useful for drier conditions as well as freezing conditions. Deficiencies are noted in the necrosis of leaf tips and edges of older leaves. As well, there may be small seed and fruit size and diminished rooting. Green shoulders on tomatoes is often attributed to potassium deficiency, while adequate potassium generally leads to improved coloring and flavor.

Along with nitrogen, potassium is very easily absorbed by plants. This ease of absorption combined with relative environmental abundance makes the excess of these two elements fairly common in vegetable production and a frequent cause of insect and disease issues, especially considering that potassium is known to increase nitrate content in plants. In analyzing potassium levels, the results of strong and weak acid extracts along with tissue analysis are often well aligned.

SODIUM

Even though sodium (Na) is considered by some as somewhat non-essential for plant growth, its role in plant metabolisms is probably just poorly understood, because sodium does have a relationship to the usage of water in plants. As well, sodium may increase the growth rate of a plant by increasing the electrical conductivity of a soil. Sodium also is known to improve the flavor of vegetables, so appropriate levels may well be worth paying attention to. Along with magnesium, sodium (Na) can impact soil structure and increase potential compaction, or a soil structure tightening. Sodium levels can be high on soil tests in arid regions yet quite low in areas of high rainfall. The interaction of sodium with other cations (Mg, Ca, K) as well as its impact on soil structure make it important to pay attention to. Sulfur is often used to reduce excessive levels of sodium in soils by increasing the solubility of sodium and thus aiding in leaching efforts.

SULFUR

Sulfur (S), like nitrogen, is a critical component of proteins and as such is often found in the organic matter of a soil. Anionic sulfur's ability to combine with the various cations makes it of use in balancing excesses of cations and in detoxifying soils by binding to toxic substances. In this sense it is often utilized for soil remediation. Adequate sulfur levels are important to keep excess nitrogen compounds from accumulating, and therefore useful in maintaining a well-balanced amino acid profile. This may be why it is known for bolstering insect and disease resistance. Sulfur compounds have traditionally been used and still are used as an active ingredient in pesticide formulations.

Brassicas and alliums are particularly high in sulfur compounds, many of which are somewhat volatile. Sulfur oxides in the air used to be a major contributor to acid rain, but there have been reductions in these particular oxides, so now less sulfur may be reaching the soil in this form. Though there are benefits in reducing the impacts of acid rain, it is conjectured that many soils and plants had become adjusted to this sulfur input and are now in a more deficient condition than when under the impact of the pollutant. As with so many elemental nutrients, too much is harmful for soils and thus plant growth. Too little is also detrimental. This is another example of why it is important for growers to understand the imbalances that are occurring and have an understanding of how to restabilize them. The most common sign of sulfur deficiency in plants is an overall yellowing of the leaves; unlike the case of nitrogen, it is the youngest leaves that yellow first. Sulfur fertilizer applications need to be approached very carefully due to its potentially antimicrobial nature.

SILICA

Silica (silicon dioxide; SiO_2) is particularly associated with the strength of exterior surfaces and is often amassed in these areas. Strengthened exteriors are very useful for insect and disease resistance. This strength of exterior manifests in many ways in the plant such as lodging resistance, root generation, erectness of leaves, and stalk flexibility. Silica is also useful in balancing potential toxicities including those from sodium (Na), chloride (Cl), aluminum (Al), nitrogen (N), iron (Fe), and manganese (Mn).

Though silica is very abundant in the soils, its forms are often very insoluble. Soils that are deficient in available silica are often high in organic matter and found in high-rainfall areas. Silica is often overlooked as an important nutritive element due to its relative abundance in the soil and the lack of increased sizing of a crop if it is utilized as a fertilizer material. Again, this material is related to the quality side of crop growth; through its ability to balance excessive nitrogen and harden exteriors, silica may actually reduce the potential size of a crop. However, it is precisely its neglect and the overapplication of sizing fertilizer materials (N, K, Ca) that make the monitoring and utilization of silica so important. Silica gives a hardened feel to a crop; its deficiency is often the opposite, which is to say a weak, lush crop, prone to lodging.

BORON

Boron's (B) roles in plant metabolism include the movement of sugars and carbohydrates, promotion of flowering and seed set, and production of lignins and cellulose. Boron is known to have a synergistic impact on calcium, which can increase cell wall strength and reduce pathogens. Boron sufficiency will aid in the movement of sugars throughout a plant. High sugar levels that have failed to be translocated from the leaves during the nighttime period are a sign of possible boron deficiency. It may be possible to measure this using a Brix meter, and testing the leaves against the roots of a plant in the early morning. Boron deficiency leads to obvious symptoms including flower buds falling off, hollow heart of roots or hollow stems, and young leaves that are rolled up.

Boron is also an anion and like the other anions it is known for being deficient in high-rainfall areas like the eastern United States. As with most of the micronutrients, boron fertilization needs to be approached very carefully. Boron fertilizers are detrimental to many microorganisms, and some plants such as legumes are easily damaged by extremely small amounts. Therefore much care needs to be taken in managing boron fertilization materials. Boron materials are often buffered with reactive carbon materials such as humates and possibly other carbon materials like molasses or, even better, compost.

IRON

Iron (Fe) is integral to many enzymes and is involved in many aspects of metabolism such as photosynthesis, nitrogen fixation by soil microbes, and nitrogen conversion. Iron's role in photosynthesis is related to magnesium and chlorophyll, and thus sufficient iron helps plants stay green and avoid magnesium-induced interveinal chlorosis. Iron's role in enzyme and nitrogen conversion makes it very critical for insect and disease resistance. Deficiencies are somewhat rare; a whitening of the leaf is the most common sign. Very high iron levels can be antagonistic to other nutritive elements such as manganese and copper.

MANGANESE

Sufficient manganese (Mn) brings on earlier maturity of fruits as well as more abundant numbers of fruits and seeds. Manganese is involved in many enzymes and reactions including photosynthesis and nitrogen metabolism. Thus manganese is related to disease resistance, likely because of its critical role in nitrogen metabolism but also for its role in lignification and creation of disease-suppressive compounds. Mn availability is reduced by a number of conditions including aerobic soils such as sands and organic soils or tilled soils. Liming and alkaline soils also will reduce Mn availability, as does high nitrate. Mn availability is increased by the opposite conditions such as acid soils, anaerobic conditions, and nitrogen in ammonium form as well as firm, untilled soils. Mn deficiency results in chlorosis of younger leaves similar to magnesium deficiency and in the chlorotic streaking of the leaves of monocots such as corn.

COPPER

Sufficient copper levels are known to enhance stalk strength and flexibility and assist in flowering and reproductive cycles, which may be due to its impact on nitrogen levels. Copper (Cu) is involved in critical enzymatic reactions in plants, many of which are particularly important for defense against diseases and insects. This may well be one of the reasons copper materials have a long history of use in agriculture to counteract plant diseases. Copper's role in plant metabolism is involved in the nitrogen-to-ammonia conversion that is critical to keeping excess nitrate from accumulating. High nitrogen levels in soils and plants may respond well to increased Cu additions. Soils high in organic matter and sandy soils are most deficient in copper. Copper deficiency symptoms include twisting of leaves and tip dieback, lodging, delayed maturity, wilting, difficulty with fruit-set, and shriveled grain.

ZINC

Zinc is involved in plant hormone synthesis and impacts the rate of maturation of seed as well as being necessary for chlorophyll synthesis. Like many of the micronutrients, zinc (Zn) is utilized in enzyme systems, possibly to a greater extent than any other micronutrient metal. Zinc is similar to copper and manganese in its utilization in fungicide formulations, and among the many zinc-based proteins are critical enzymes for synthesis of compounds that defend plants against insect and disease. This brings up whether the success of fungicides based on copper, manganese, and zinc is due to their ability to destroy pathogens, to enhance a plant's defensive compounds, or to both. Common deficiency symptoms include small leaves and slow spring growth, cracking and splitting conditions, poor seed set, and chlorotic mottling of leaves.

CHLORIDE

Chloride (Cl) is known to help with water transfer through cell walls and thus to be of assistance in drought conditions. Chloride also has a role in photosynthesis and may increase manganese availability in the soil. Chloride is an anion of essential nature to plant growth yet its monitoring by soil test, and its use as a fertilizer, are somewhat neglected due to its relative abundance in arid regions, as well as unintended inclusion in common fertilizer materials like potassium chloride. However, in high-rainfall areas intentional chloride fertilization can lend important improvement to insect and disease resistance, especially in today's stressful conditions. Chloride is involved in the nitrogen cycle by inhibiting nitrification, which is to say the conversion of ammonium NH_4 to nitrate NO_3. This again may be of benefit by limiting the accumulation of excess nitrate. Chloride is often attached to a nutritive cationic element such as potassium or sodium in fertilizers, and it is conjectured that some of the positive impacts attributed to the cations in such fertilizers may well have been the result of increasing chloride in the system.

MOLYBDENUM

Molybdenum (Mo) is highly related to nitrogen activity in plants and soils. Mo is critical for the enzymatic reaction of nitrate reduction in the plant and is essential for soil microbial fixation of atmospheric nitrogen. Mo is thus known to greatly assist in the reduction of nitrate levels in leaves and thereby assist in disease resistance. Mo is also known to increase nitrogen levels in soils so that plants can access more of this nutrient if necessary. It is regarded as sufficient in extremely small amounts. It must be remembered, however, that *sufficient* levels and *optimal* levels may well be very different. Molybdenum appears critical for a highly functioning nitrogen system and its usage in fertilizers has greatly increased recently, especially small amounts in foliar preparations. Mo deficiencies may be quite extensive yet are not well diagnosed. Some labs do not include Mo as part of their standard testing, and soil levels may be so low that it may be difficult to assess what is an optimum amount. Tissue analysis can give a clearer picture of how much molybdenum is getting into a plant, yet optimal tissue levels are also not well established. An on-farm test to determine whether molybdenum is deficient is to split open the rhizobium nodules attached to the roots of legumes and examine the interior coloration. Deficiency is likely if the color is white or greenish; sufficiency is more likely if a red-brown coloration is present.

OTHER MICRONUTRIENTS

A few of the many other nutritive elements that benefit plants include cobalt (Co), nickel (Ni), selenium (Se), and iodine (I). Some have been declared essential to plant growth, meaning that a plant cannot grow to fruition without their presence, though many more are beneficial in various regards to the functioning of a plant as well. It is possible that almost all the chemical elements have at least some small role to play in soil and plant systems. Nature has provided a balance to their availability and function in the past that is likely quite disturbed now. As soils have been particularly damaged in this balance, it may well be worthwhile to bring in materials from relatively clean areas of the ocean where more of these beneficial trace nutritive materials may still be circulating in the living systems and introduce them onto cropland. Common sources would be seawater, seaweeds, or fish fertilizers. These materials may well contain minute amounts of nutritive elements whose function may not be understood by science at this time. Another potential source of trace minerals that may be appropriate for a given soil is the crushed bedrock from below it. Local quarries often offer these rock dusts at very affordable rates and these materials may be of appropriate balance for a soil that is derived from this rock. Other commercially available rock dusts may also contain some of these trace minerals to a greater or lesser degree.

Many elements that are considered nutritive may well be poisonous if applied in too large a dose. Often soils are lacking in the most hazardous of these materials, such as sulfur (S), boron (B), and copper (Cu). Great care is required in the usage of fertilizer materials that contain these and other micronutrients, both in terms of soil biological exposures and human applicator exposure. Some soil and tissue labs offer testing for substances considered toxic to living systems, such as lead and aluminum. While it is very true that the overuse of poisonous heavy metals has resulted in their accumulation to a detrimental degree in our soils, it should not be forgotten than even these metals may have a small role to play that would be beneficial if properly balanced. Aluminum is a prime example of this: It has been shown to be essential for a number of plant species, yet it is quite well known for toxicity in soil systems and human health.

CHAPTER 9

Fertilization Materials and Methods

V egetable growers have traditionally brought large volumes of nutritive materials onto their lands, often at yearly rates approaching 100 tons to the acre (247 metric tons per ha) or more. This is a task that can require a significant amount of effort and is often mechanized. The addition of large volumes of composts and mulches remains the basis of vegetable fertilization efforts, though often this can be supplemented by modern fertilization materials, which may be quite concentrated. Some of the micronutrient materials discussed in this chapter, for example, can supply nutrient needs through very small additions of a few pounds per acre, or even less.

The general objective of most fertilization is the development of improved soil and plant function. However, it is critical that the overarching four elemental conditions have been properly attended to or fertilization efforts will be largely wasted. Attention to the basic principle of keeping the soil covered at all times with both decaying residues and mulches as well as an adequate plant canopy is also primary. Fertilization efforts will impact the balance of growth and reproductive forces as well as characteristics of soils, including aggregation and biological conditions, warmth, water/air, and drainage characteristics. The impacts of fertilization need to be observed thoroughly in concert with all other agricultural techniques because they have an interrelationship that affects the balance conditions. Tillage is a primary consideration in this regard, as well as irrigation and drainage projects.

Growth and decay are a continuous loop; one feeds the other. This loop is intimately tied to the soil's biological functioning, and the growth and decay cycle is a primary dynamic growers manage in order to enhance soil fertility. As vegetable growers, our goal is to increase this cycle to higher levels of activity, and thus bring about conditions of better crop growth. Indeed, enhancing the growth and decay process is the primary goal of fertilization. Observing how fertilization impacts this dynamic is a critical component of evaluating the appropriateness of fertilization efforts, and thus trials, testing, and guidance are particularly important to help develop our ability to apply the correct fertilization materials at the correct time.

The trialing of fertilization materials is particularly important, often achieved with the simple trial of not applying a fertilizer to a small area as a control for comparison. It is also relatively easy to set up trials of one fertilizer versus another, with repetitions at different times and conditions being even more useful. Supplementing direct observation with soil and tissue analysis of the different treatments may be worthwhile as well. Testing with a refractometer, pH meter, and electrical conductivity meter may also be helpful to compare a fertilized crop with an unfertilized crop and note any benefits or detriments as a result of the application.

There are essentially two approaches to fertilization: the fertilization of soils and the direct fertilization of a crop. The fertilization of soils is the addition of materials so that the soil provides improved conditions and nutrients for crop growth. In the case of direct fertilization of a crop, the goal is to deliver materials directly to the crops, with little or no soil interaction. Foliar fertilization is an example. Often the two approaches overlap, because foliar-applied materials influence the composition of a crop's root exudates and thus the soil's fertility. On the other hand some of the nutrients in a soil-applied soluble fertilizer may well be microbially processed before being absorbed by plant roots, while other nutrients are directly absorbed by roots. If the crop needs immediate nutrient availability because an imbalance causes serious concerns for quality or yield, applying a highly available fertilizer material as a foliar feed or a liquid side-dressing may well be the best option. However, it is desirable to direct most fertilization efforts toward long-term soil fertility gains by adding materials meant for microbial processing. These materials, if appropriate, increase the living force and mass of a soil system and thus enhance the soil's ability to provide a steady, complex, balanced nutrient release to the growing crop.

Many modern fertilization materials are soluble inorganic materials that readily dissolve in water. These materials are easy to apply through water-based systems such as irrigation. They are also easily assimilated into plants and soils. They produce a quick, dramatic impact on both crop growth and soil life, but they can be readily lost from the soil system through leaching and volatilization. Many potentially useful inorganic fertilizers are toxic to life in their concentrated forms. This includes toxicity to the grower—nitrogen poisoning from handling fertilizers being a prime example. (Many soluble fertilizer materials are readily absorbed through the skin.) Caution is definitely in order: slow, extensive trialing of small amounts of materials until their benefit is demonstrated.

Many mineral materials are not as soluble, but are brought by soil life into forms that plants can assimilate. These include ground rock materials such as limestone, talc, gypsum, rock phosphate, wollastonite (calcium silicate), and so on. These materials are often of a slower, gentler disposition toward soil life and can be very useful in the slow, careful guidance of soil conditions. They are often applied in much larger volumes than soluble materials. Caution is advisable when handling these materials, too, because insoluble ground rock particles are quite dusty and can cause potentially serious damage to the lungs.

Recall here that especially when writing about fertilizers, the term *organic* is used in the sense of "containing carbon," not in terms of materials acceptable for use on certified organic farms. Both insoluble and soluble fertilization materials are often inorganic (non-carbon containing), and benefit from being mixed with organic materials both to buffer their potential of damage and to begin their assimilation into living systems. One way to accomplish this is to mix into the composting system inorganic materials such as rock dust, lime, rock phosphate, and talc. Another is to add carbon-containing materials to fertilizer mixtures. Common choices of organic additions to fertilizer mixes are molasses, humic compounds from humate deposits, charcoal, compost teas, or compost. The buffering capacity of these carbon materials prevents the reactive components of fertilizer materials from overreacting chemically with one another as well as providing protection to any living component of the fertilizer. Again, many inorganic fertilizers are toxic to life when they are concentrated.

Many micronutrients are poisonous even in small amounts; great caution is in order when working with them in concentrated forms! Unfortunately many of the more dangerous micronutrients such as copper, boron, zinc, and molybdenum are the ones of most benefit for damaged agricultural environments. The saying "a little poison makes you stronger" may be accurate in these cases. These micronutrients often appear to be lost to the living systems of our soils at this time. However, living organisms utilize the macronutrients in much greater volumes, and these nutrients are worthy of the grower's primary attention. For example, it will not matter how much zinc, boron, or molybdenum is available for a crop if the soil's carbon to nitrogen ratio is way out of balance!

Fertilizer Application Timing

In general, it is best to apply fertilizer materials slowly and steadily to avoid disruption of living soil systems. This applies to both soluble and insoluble materials. Though it may take more effort to make multiple applications than a single large one, the benefits of a gentle approach will usually be worth that effort. Not only is there less disturbance to the living soil, but with multiple applications a grower is able to more precisely deliver the appropriate nutrients for the growth of a crop.

The three primary times to apply fertilizers are preplant fertilization, side-dressing a growing crop, and foliar application. Preplant fertilizer is often broadcast over the entire growing area, though it can also be banded alongside the future crop row. Broadcast preplant fertilization is an opportunity to move overall field balance and nutrient levels in a particular direction; compost application is a common example of this. Preplant fertilization is also a time to provide large volumes of nutritive substances to the soil life to increase their functioning and to apply less-soluble fertilization materials that need to be biologically digested, like the rock dusts. The bulk of fertilizer materials are applied preplant. For vegetable crops, 30 tons or more of compost to the acre (74 metric tons per ha) preplant is relatively common. Preplant is generally not the appropriate time to apply much soluble fertilizer to direct-seeded crops, because the crops' nutrient demands will be negligible for a significant amount of time following germination. For transplants, applying soluble preplant fertilizers is more common.

It is important not to overfeed young seedlings, because they are very prone to weakness when nutrient levels are excessive. The innate nutrient availability in biologically active soil combined with the vigor from well-grown seed is usually sufficient for the early growth of a crop. Later in its development, a crop's nutrient needs are exponentially greater. That is the time to side-dress to help supply the nutritional needs for that particular crop at that specific stage of its growth. (Refer to table 2.2 for general nutrient guidelines for different crop growth periods.) Fertilizer materials used for side-dressing are often soluble or highly biologically activated. The volume applied is far less than the tonnage of preplant applications of compost and insoluble fertilizer materials, yet much more than the amount applied in foliar applications. Side-dressing is an opportunity to guide an individual crop to a specific goal, such as sizing of the leaves and delaying flowering, moving the crop into early flowering, or sizing fruits. Side-dressing materials may be either liquid or solid. The solid side-dressing materials we use are all compost-based. Essentially, we build compost piles specifically for the various periods of growth. These piles are usually relatively small, contain many added minerals, and are of short duration. For more on creating such specialized piles, see chapter 10. A crop can be side-dressed with liquids or solids several times during its development, beginning with a small application onto or next to seedlings or transplants and continuing until maturity.

Materials for foliar application are usually soluble and almost always applied as a liquid. An example of solid foliar application is kaolin clay, which is applied to harden and protect foliage against damage by some insects. Liquefied foliar materials are often applied in very small volumes. They supply small amounts of nutrients in highly available forms, especially trace minerals and beneficial biology. Foliars can be effective to fine-tune efforts to guide a crop to a specific goal such as sizing or flowering. Such an application often coincides with side-dressing, which provides the main impetus toward the goal. This could be the use of a high-phosphorus side-dress fertilizer combined with magnesium chloride and vinegar in the foliar to initiate early heavy fruit-set in tomatoes, for example. Foliars are of great assistance in enhancing disease and insect resistance as well as increasing pigmentation and guiding a crop to maturity. Many effective foliar materials are of a living nature and would be damaged by the addition of soluble mineral fertilizers. If a soluble fertilizer material needs to be applied, it may well be best to have a separate application, or at least be thoroughly buffered by carbon materials. When foliar spraying vegetable crops, it is particularly important to avoid applying materials that could disflavor a crop, such as liquid fish fertilizer, or mineral materials that are inappropriate for consumption too close to harvest. For our crop production, we find it relatively easy to keep the foliar program in the realm of fully palatable and non-toxic materials.

Mineral and Inorganic Fertilizers

Compost is a primary fertilizer, particularly for preplant usage. (See chapter 10 for details of the preparation and application of compost, including addition of minerals in specific composts.) Below is a description of mineral and inorganic fertilizer materials that are both blended into compost materials or applied via side-dressing.

Understanding a little basic chemistry helps in determining the best use of these materials. Most inorganic fertilization materials are a combination of positively charged ions and negatively charged ions that form a relatively stable compound called a salt. Some salts are more soluble than others. A fertilizer may also be in ionic form, meaning it is chemically reactive and not combined with materials of its opposite charge and thus is not stabilized. Table 9.1 lists the common cations and anions utilized in fertilizers.

Here is one example: Calcium can combine with nitrate to form calcium nitrate, a common synthesized fertilizer utilized in vegetable production. Calcium can also combine with carbonate to form limestone, with phosphate to form rock phosphate, with sulfate to form calcium sulfate (gypsum), with silicate to form wollastonite, with borate to form calcium borate (a boron fertilizer), or with chloride to form calcium chloride (road salt). Sodium can take the form of sodium nitrate (Chilean nitrate), sodium carbonate (washing soda), sodium phosphates (detergent), sodium sulfate (glauber's salt), sodium silicate (water glass), sodium borate (Solubor), and sodium chloride (table salt). Many of these salts are naturally

TABLE 9.1. Common Cations and Anions in Fertilizers

Cations	
NAME	**FORMULA**
Ammonium	NH_4^+
Calcium	Ca^{2+}
Magnesium	Mg^{2+}
Potassium	K^+
Zinc	Zn^{2+}
Copper	Cu^{2+}
Iron	Fe^{2+}
Manganese	Mn^{2+}
Sodium	Na^+
Anions	
NAME	**FORMULA**
Sulfate	$(SO_4)^{2-}$
Carbonate	$(CO_3)^{2-}$
Phosphate	$(H2PO_4)^-$ and other forms
Silicate	$(SiO_3)^{2-}$
Chloride	Cl^-
Borate	$(BO_3)^-$
Nitrate	$(NO_3)^-$

occurring minerals that are simply mined, ground to varying degrees, and delivered as fertilizer materials. The more finely the minerals are ground, the more quickly they can be acted upon by the soil biology. Some of these minerals can also be created by synthesis in fertilizer factories, often through the use of heat. Uncombined ionic fertilizer materials can be created through heating a mineral, as in the case of turning calcium carbonate ($CaCO_3$) into calcium oxide (CaO)—a cationic form also called burnt lime—and carbon dioxide (CO_2). The calcium oxide can then be combined with an acid such as nitric acid to form calcium nitrate or with sulfuric acid to form calcium sulfate (synthesized gypsum). This is how many of the micronutrient sulfate fertilizers are created.

The cationic oxides have a long history of usage in crop production; wood ash is an example. These substances are very reactive, however, and potentially hazardous to soil life, so a cautious approach to their usage is in order. This is also true of the acids formed by anionic substances, such as sulfuric and phosphoric acid. Many fertilizer materials are synthesized through the action of heating or chemical reaction. Some of these are approved for use in USDA certified organic farming operations and some are not. Approval is somewhat variable at this time. Many fertilizers obtained by heat are acceptable such as wood ash or charcoal, yet burnt limestone is not. Sulfuric and phosphoric acids cannot be utilized directly, but are utilized in liquid fish fertilizers as a stabilizing agent and nutrient source. Many acceptable micronutrients are synthesized through heating of ores and subsequent addition of sulfuric acid to produce the suitable forms of zinc sulfate, copper sulfate, and so on. Many synthesized fertilizer materials need to be carefully trialed by growers before adoption into a fertilizer program, as they can be hazardous to soil life if misapplied, but it is up to individual growers to determine their potential usefulness in their growing conditions. This chapter covers both synthesized and naturally occurring fertilizers that can be useful in vegetable production.

Nitrogen compounds are rarely found in terrestrial mineral form that can be mined, and this is why nitrogen is the most commonly synthesized fertilizer material. The exception is a very large deposit of sodium nitrate (Chilean nitrate) located in a mountainous desert in Chile. Nitrogen is present in relative abundance in organic materials, as described later in this chapter. The use of organic nitrogen materials from the realm of life is generally preferable to nitrogen salt fertilizers. Nitrogen salt materials are very soluble and can cause tremendous impacts on growing conditions, often not for the better. Nitrogen from the living realm has much less potential to cause damage to soil life, and thus use of such materials is generally preferable to use of nitrogen salt fertilizers. Organic nitrogen often can be reassimilated by life in a more complex form than a simple salt mineral—as amino acids, for example. When a growing plant can assimilate an existing amino acid rather than having to synthesize it, this is beneficial for the conservation of energy in the plant.

Some common fertilizer materials are sulfate forms of cations, including calcium sulfate (gypsum), magnesium sulfate (Epsom salt), and sulfate of potash, all relatively abundant naturally formed minerals. The sulfates are relatively soluble and gentle on living systems, hence their popularity as a fertilizer material. Although many of the trace mineral sulfate fertilizers exist in natural forms, the products sold as fertilizers are synthesized in factories using sulfuric acid and an appropriate ore, which generally yields an acceptable fertilizer material.

Carbonates occur widely in naturally formed minerals. Calcium and magnesium carbonate are the most frequently used in agriculture. The carbonates are less soluble than the sulfates, yet are still accessible for use by living systems. Calcium and magnesium carbonate combined is called dolomitic limestone; limestone with little or no magnesium is called calcitic limestone. When these limestones are heated to a high temperature, burnt lime (calcium oxide and magnesium oxide) is formed. There are other naturally occurring carbonates, but their availability as fertilization materials is limited.

Of the phosphates, only calcium phosphate (rock phosphate) is a common naturally occurring fertilizer material. Vegetable growers do make use of finely ground soft rock phosphate. But because rock phosphate is somewhat unavailable to crops directly, it does best in soils where biological, particularly fungal, activity is high and thus can make the material accessible. The approach we use at Tobacco Road Farm is to incorporate fine-ground soft rock phosphate into an active, fungal-rich compost pile, allowing the high level of biological activity to work upon the mineral. Soft rock phosphate can be mixed into water, and the finer particles will remain suspended for some length of time. The suspension can be applied as a liquid side-dressing as long as it isn't left sitting too long. Phosphoric acid is commonly used to stabilize liquid fish fertilizer and increases its phosphorus content.

More soluble forms of phosphate fertilizer are manufactured from rock phosphate. One such form is phosphoric acid, which can be synthesized further to form superphosphate. The soluble phosphorus materials are notorious for quickly reverting to insoluble forms such as calcium phosphate in the soil. Highly functioning soil biology is needed to ensure steady phosphorus release. If this biological condition is not in place, soluble phosphorus materials can be steadily applied in small doses to keep a supply available to crops. Some growers apply phosphoric-acid–stabilized fish fertilizer steadily fed through a drip irrigation system.

Silicate fertilizer materials in common use are calcium silicate (wollastonite and diatomaceous earth), magnesium silicate (talc), and potassium silicate (greensand), as well as ground rock from quarries, such as trap rock and granite. Many cationic elements also occur as silicates. Like limestone, the silicates can be combined into calcium-magnesium silicates (such as Azomite) and other mixtures. These are naturally occurring mined materials. The silicates are not used as frequently as carbonates or sulfates. However, with fine grinding and highly biologically active conditions they can be very useful materials. Supplying available silica for plant growth can greatly assist plant strength and flexibility, thus improving pest and disease resistance. Both calcium silicate and magnesium silicate can assist with balancing high nitrogen and potassium levels. Adding silicate fertilizers to a composting system may work best to help bring them into availability. The silicates do not disrupt composting dynamics the way some soluble salt materials can, and clay-humic complexes can form on finely ground silicates during composting. Quarry dusts are a common inexpensive source of silicates, which are abundant in the crushings of granite or trap rock (basalt), as well as many other types of quarried rocks.

Chloride fertilizer materials are calcium chloride (road salt), magnesium chloride (seawater extracts), potassium chloride (muriate of potash), and sodium chloride (table salt). Although chloride is a necessary micronutrient, chloride fertilizer materials are highly soluble salts capable of disrupting and damaging biological systems. Soils, especially highly leached ones, may benefit from the addition of chloride, but chloride can build to excessive level in soils, so care is necessary to avoid overapplying. Calcium chloride and magnesium chloride often are added in small amounts to foliar sprays. Commercial fertilizer blends often contain potassium chloride; cautious use of such materials is advisable, because both potassium and chloride can easily build to excessive levels. Sodium chloride, which is common table salt and the primary ingredient in sea salt, may also be a useful fertilizer in high-rainfall areas where soils are leached of their soluble salts. Both sodium and chloride may be low in these soils. A preferable source of sodium chloride is seawater itself or seawater that has been dried and solidified into a mineral salt with other naturally occurring minerals still present. (For more about seawater, see "Foliar for Fruit Ripening and Disease Prevention" on page 156.) There are similar mineral salt deposits of terrestrial origin. These salts contain a variety of potentially useful trace minerals in small amounts, often in quite available forms as well as in a reasonable balance for living systems. Often in the manufacture of commercial sea salt, sodium chloride salt is extracted from the seawater for sale as a purified product. This leaves the remaining salts behind in a slurry. This slurry material, which is high in magnesium chloride along with other trace mineral salts, is sold as fertilizer (Sea-Crop).

Other anionic elements are also combined into salt fertilizers; the most common include sodium borate (Solubor), calcium borate, sodium molybdate, and various iodates. These materials are used in very small quantities for very specific applications. Again, caution and observance of results are very important because of the potential damage to living systems. Borates can be particularly damaging to legumes and other plants even at extremely low doses. Introducing these materials through the composting system may be most appropriate, but some growers carefully apply these materials in very small doses through foliar or side-dressing applications, often with a carbonaceous buffering material added.

Fertilizers from the Realm of Life

Many common fertilizers for vegetable crops are organic (carbon-containing) materials from the realm of life. These materials are often composted before applying, but some of them may also be applied uncomposted as side-dressings or foliar sprays. Being originally from living organisms, these materials often contain a more balanced nutrient profile than mineral materials. They are not primarily composed of soluble salts yet they are generally more available for life to decompose and reassimilate than many minerals. Because the realm of life has become so polluted in our time, organic materials are often prone to chemical contamination. Unlike mineral fertilizers, many organic (carbon containing) fertilizer materials can be directly gathered by the grower from the local environment, and thus better assessed for qualities and possible contaminants.

ANIMAL-BASED FERTILIZERS

Many of the carbon-based fertilizer materials come from animals. These include blood meal, bonemeal, fish meal and liquids, guano and manures, and milk. Many of these materials are relatively high in nitrogen, and stabilizing them through composting with carbon materials before application, even for a short period, is often of benefit.

BLOOD MEAL. Commercial blood meal is collected from slaughterhouses and may be contaminated with chemicals used in the production and slaughter of animals. Blood may be best when collected from on-farm slaughter. Blood is very invigorating to soils but may easily be used to excess.

BONEMEAL. Bonemeal is largely composed of calcium phosphate. It is similar to rock phosphate but, depending on the degree of grinding of the materials, may be more accessible to soil life. Bone meal is commercially available, but bones can also be collected and processed on the farm, especially from composted carcasses where the bones fail to break down. To process bones it is often easiest to charcoal them first. We do this by cutting open part of the lid of a 1-gallon (3.8 L) steel container (we use an olive oil can), filling it with bones, and setting it into the woodstove firebox to heat. We pull the bones out before they turn to ash and allow them to cool. The charred bones are then easily ground up; for our small batches we use a meat grinder. Larger batches can be pounded with tools such as a piece of a tree limb about 5 inches (13 cm) in diameter.

This basic process of heating a material to bring it into an alkaline state is a common process in large-scale fertilizer manufacturing. Before the development of modern grinding equipment, this process was utilized for lime production because it made the limestone much easier to pulverize. The burning of limestone creates calcium oxide (CaO), which takes on hydrogen when exposed to moisture and becomes calcium hydroxide ($CaOH$), also called hydrated lime, which is still quite alkaline. The oxides or hydroxides of calcium or magnesium from burnt limestone can also be reacted with vinegar (acetic acid) for improved availability. The reaction of acetic acid with these calcium-containing alkalized materials forms a significant amount of calcium acetate, a highly available calcium fertilizer.

FISH FERTILIZERS. Fish meal and liquid fish fertilizers are often produced from ocean-dwelling fish, which may have better access to a broad array of trace minerals than land-based animals. These fertilizers are commonly produced from the carcasses of fish processed for human consumption, and they are relatively high in nitrogen-rich protein. Some methods of processing fish for fertilizer employ chemicals that may be of detrimental impact. Many fish liquids are "stabilized" with sulfuric or phosphoric acids, which makes them quite acidic; care in handling is in order when using such products. Sulfuric and phosphoric acids are also fertilizers of potential benefit or detriment depending on the individual situation. Dried fish meals may contain different chemicals utilized in processing, some of which may not be beneficial to living soil systems.

Since our farm is in close proximity to the ocean, we can collect fish carcasses from fish markets and processors where the fish are being filleted. The carcasses can be composted to provide an initially smelly but very nutrient-rich compost, or processed into farm-made liquid fish fertilizer. To brew the liquid fertilizer, we first combine the fish carcasses with unprocessed brown sugar at a 1:1 ratio by weight in various containers. A little compost or IMO (described throughout chapter 11) is spread over the fish, and then some straw to top it off. The container is covered with window screening (to keep flies out) and allowed to sit for six months or more. This is an odoriferous process. We strain the fish/sugar mixture through a plastic mesh basket into 5-gallon (20 L) buckets. The nitrogen content is obvious from the strong ammonia smell upon straining. This farm-made fish liquid is free from chemical additives and is full of living organisms that aid in its assimilation into the living system. This fertilizer is generally too smelly to use for foliar applications, so we apply it as a side-dressing. Because it is nitrogen-rich, it is particularly helpful to stimulate early growth of plants and particularly helpful for transplants.

MANURES. Manures are the animal-based materials most used in vegetable growing. These can be in the form of bagged guano or dried pelletized chicken manures but most frequently are simply sourced in raw form locally. Manures can be, and on many farms often are, directly applied to a growing area but in general greatly benefit from going through a composting process first. Manure sources are often contaminated with chemicals detrimental to crop growth, including antibiotics, dewormers, hormones, and pesticides (from animal feeds as well as materials applied directly to and in the animals). Some of these materials are damaging to living systems. They are not called antibiotics and dewormers for nothing! Of particular concern are the herbicides, some of which are extremely persistent, used in the growing of grasses and grains for feed. Some herbicides do not readily decompose, and as a result manure can contain residues, which can be damaging to broad-leaved (non-grass) vegetables. Soils can detoxify themselves to some degree, but often these contaminants can overwhelm the soil's capacity. It is always best to investigate the source of manure and potential for contaminants before usage.

While at Tobacco Road Farm all manures are incorporated into the composting system, farms may have reason to directly apply manures to soils. Uncomposted manures are usually applied preplant, often well before the crop planting date. A common approach is to manure a vegetable field in the fall, before or while a cover crop is growing, with vegetables to follow in the spring. Applying undecomposed manures to soil is sometimes referred to as sheet composting, especially when other organic materials are spread besides just manure. Applying undecomposed materials and allowing them to decompose either on top of the soil or in the soil (by tilling them in) saves on the effort of building compost piles but can disturb soil biological conditions, especially in the case of incorporation with tillage. A frequently successful approach home gardeners use is to apply undecomposed compost and household scraps to garden beds in the fall, then cover that layer with chopped leaves and grass clippings. By the following spring, the compost and scraps are often fully composted, and gardeners can transplant through the surface leaf mulch, or rake the mulch aside to sow seeds.

The nutrient balance of a manure source is highly related to the animals' diet. Manure from grain-fed animals, particularly poultry, often has high levels of N and P. Manures from laying hens often have a high calcium content because of the calcium added to the feed. Manures from pastured animals reflect the nutrient condition of those pastures. With their extensive digestive systems, ruminants often add beneficial biology to their manure, which assists in its reassimilation into living systems. Often the most practical, least contaminated manure in our region comes from grass-fed cattle that graze where the pastures and hay lands are not treated with herbicides. These animals are generally quite healthy and unmedicated, with few chemical additives in their feeds. A hay/manure mixture collected from areas where these animals were fed hay in winter makes an excellent compost addition.

RAW MILK. Raw milk is another of our favorite fertilizer materials. Milk is a tremendous growth force material. Nature designed it to nourish young animals, but it also does well by young plants. It is generally high in protein (and thus nitrogen) and calcium with an excellent balance of other nutrients in a highly available form. It can be diluted to varying degrees for use as a preplant, side-dressed, or foliar fertilizer. Raw milk is relatively stable and does not go bad quickly, unlike pasteurized milk products. The cream forms a barrier to decay if allowed to seal over at the top of the container. However, the cream portion can clog liquid fertilizer application systems, so milk may need to be strained before application. Milk may be available from dairies,

dairies, particularly in periods of overproduction, or when gone out-of-date or otherwise too old for sale, for little or no cost. For foliar applications, the benefits of purchasing raw milk off the shelf may outweigh the cost.

PLANT-BASED FERTILIZERS

Organic fertilizers from the realm of plant life often make up the bulk of ingredients in a composting system. Although composting of materials generally best prepares them for use in agricultural systems, some can also be utilized in an uncomposted state.

SEED MEALS. Seed meals are among the most common plant materials utilized uncomposted. These are usually the leftovers from oil extraction, brewing, bakeries, or other food processing. Being grains, they are generally high in N and P. Grain hulls and brans from grain grinding are also sometimes utilized. The hulls, being from the exteriors of the grain, are higher in C and Si. These materials are often side-dressed or top-dressed by being spread around plants.

SEAWEEDS. Like fish materials, seaweeds can be very useful in vegetable growing due to the high levels of trace elements from the ocean environment. Seaweeds can be dried or liquefied. Both seem to contain fewer chemical additives or contaminants than many other organic fertilizers. Seaweeds can certainly be gathered by growers, though some ocean areas may have chemical contamination present. Seaweeds do not have a high N-P-K analysis but offer both important trace elements in a balance established by the organisms, and also various naturally growth-enhancing hormones such as gibberellin. If it is possible to gather a large quantity of seaweed, it can make an excellent compost addition. Smaller quantities can be added to side-dressing formulas. Commercial liquefied seaweed often makes an excellent addition to most foliar sprays.

MOLASSES. Molasses is a relatively inexpensive fertilizer. It and other sugars provide a quick food source for bacteria, and thus can be useful in mixed fertilizers to stimulate microbial activity that will act upon and release nutrients from other fertilizer materials in the blend. Being soluble and largely composed of carbon, molasses can also help buffer a liquid fertilizer blend and prevent materials in the blend from reacting with one another. Molasses is particularly useful in side-dress formulations where quick release of fertilizer materials is important.

CHARCOAL AND HUMATES. Natural charcoal and humates are carbonaceous materials that are useful for buffering fertilizer blends. Both are essentially long chains of carbon molecules, providing chemical bonding sites to nutrients while keeping them in an available form. Charcoal is produced by burning carbonaceous materials in a low-oxygen environment. Humates are the result of low-oxygen decomposition of organic materials over lengthy periods of time. They are basically the precursor of coal. Charcoal can certainly be made on the farm. Humates are available commercially. These materials, particularly liquefied humates, can be useful for buffering liquid fertilizer mixtures. A similar buffering material can be made on-farm by the liquid extraction of humus-containing composts or vermicomposts. Charcoal or humates are useful for the buffering of dry fertilizer blends, although compost can serve this role, too.

VINEGAR. As a naturally occurring carbon-containing acid, vinegar can be useful to move plants toward flowering and reproduction as well as to harden them, and to stimulate insect and disease resistance. Many materials can be made into vinegar, and vinegars are available in different strengths of acidity. The most popular commercially available food-grade type fertilizer is apple cider vinegar. Most vinegar can be produced by the grower in a relatively easy process, although it is often difficult to get higher levels of acidity. Vinegars are often used in side-dressing formulas or foliar preparations.

Applying Fertilizer

Most vegetable growers apply solid materials preplanting, and often in large volumes, as noted earlier. Solid materials like compost or manure can be spread with manure spreaders, dump trucks, or wheelbarrows. A loader tractor is of great assistance for this task.

Other mineral fertilizers can be broadcast over fields with the use of drop spreaders or spin spreaders, or by hand with buckets. When spreading dry mineral materials, it is important to avoid breathing in the dusts. Spin spreaders can generate a lot of dust when spreading dry mineral materials. In these cases, it is best to wear protective gear to avoid breathing in the dust.

Adding a moist material like compost to the mixture can cut down on dust when spreading dry materials. For our limited dry fertilizer applications, we mix any dry preplant fertilizers in a cement mixer with a little compost added to cut the dust and then use buckets and scoops to apply it. For the most part, though, we do not apply dry mineral materials directly preplant; we add most dry minerals to compost piles.

Liquid fertilizers can be applied preplant through irrigation, sprayers, or watering cans and often find the most usage as side-dressings and foliar sprays. To apply liquid fertilizers through an irrigation system, a tank for liquid fertilizer can be installed into the water line as shown in figure 9.1. The pipe

FIGURE 9.1. The line from the fertilizer tank can be inserted into our irrigation system's suction line using the quick-connection fittings. The valve on the fertilizer line adjusts the flow.

on the suction side of the pump is installed with a T-fitting and a valve on the line to the tank to regulate flow into the suction line. This requires an isolated water source not utilized for drinking and a foot valve on the end of the water suction line to prevent backflow of any leftover fertilizer to the water source. On the output side, sprinklers or hoses can deliver the material, or it can be filtered and run through a drip irrigation system. We use full-flow handheld fan-pattern sprayer nozzles attached to a 1-inch (2.5 cm) diameter hose for maximum output.

Specialized fertilizer side-dresser attachments mounted on tractors may work well under some conditions, but generally we apply solid side-dressed materials by hand. When applying by hand, use a hoe to make a furrow alongside the crop row, sprinkle in the material directly from a bucket, and then use the hoe to cover the furrow. Another approach for mulched vegetables is to rake the mulch aside, spread the material on the exposed soil, then re-cover with the mulch. This is very effective; feeder roots of crops quickly grow into this nutrient source, especially if the materials are watered in. Side-dressing can begin at seeding or transplanting and continue through the growth periods

FIGURE 9.2. This liquid side-dress is being applied with a fan spray hose end to the base of a crop of celeriac.

as needed; need being determined by careful observation of growth and also possibly tissue testing.

Developing Fertilizer Formulas

We have developed our own formulas for side-dressing and foliar application. They are provided as examples to give insight into how materials can be combined and in what relative volumes. These recipes have been developed for the particular soils and environmental conditions of Tobacco Road Farm: high organic matter, high cation exchange capacity, valley loam soils under the influence of intense pollution, and a generally degraded regional ecosystem subjected to climate modifications. Because of our growing situation, our recipes are relatively complex, and we frequently alter the ingredients based on our assessment of specific crop needs, or because of field conditions or availability of materials. Simpler versions may well work in other situations.

Our formulas include different ingredients for different growth stages of crops, and the ingredients lists are offered simply to help stimulate the grower's thought process. We use stage II materials the most, because in our situation there is often more than enough growth force to provide for initial leaf growth (stage I).

However we might still utilize a stage I recipe for cabbage head sizing, for example. A recipe for a fruit-sizing (stage III) solid material to apply by side-dressing is included, but on our farm plants are often so large and difficult to access at that stage of growth that we frequently choose to rely on foliar fertilization instead.

For *all* of the following liquid blends, we combine the ingredients in 100 gallons (380 L) of water in the fertilizer tank. The flow from the fertilizer tank is adjusted to further mix this with irrigation water at a ratio of about 1:20. This makes this mixture quite dilute.

Seed-Starting Liquid

We often use a liquid blend formula when we are irrigating newly seeded beds. In this blend we choose to use a relatively weak liquid seaweed product; if we are using a concentrated form, we cut back on the amount added. We include fermented plant extracts if they are available; carrot and beet extracts are our favorite for seed treatment. (See "Foliar for Fruit Ripening and Disease Prevention" on page 156 for instructions on making fermented plant extracts.) To gather vermicompost extract, we water the vermicompost bins about an hour or so in advance, collect the runoff, and immediately utilize it. Liquefied compost extract is a good substitute that is easy to make. We put about 1 gallon of compost in a 5-gallon (20 L) bucket and fill it with water, then stir vigorously, strain, and use immediately.

- 20 fluid ounces (600 ml) liquid seaweed
- 20 fluid ounces (600 ml) apple cider vinegar
- 10 fluid ounces (300 ml) fermented plant juices (FJP)
- 20 fluid ounces (600 ml) carrot/beet plant extract (a type of FJP)
- 20 fluid ounces (600 ml) Epsom salt (magnesium sulfate)
- 5 gallons (20 L) seawater (or 40 fluid ounces agricultural sea salt)
- 5 gallons (20 L) liquid vermicompost extract
- Small amounts of other trace elements such as molybdenum, boron, and sulfur (if deemed necessary)

The quantities listed here will make enough liquid to cover roughly 5,000 square feet (465 sq m). We apply by hand via high-volume hoses with flat fan-spray-pattern nozzles attached, covering the entire surface of newly seeded beds, at a rate appropriate to the soil condition, in order to bring the seed to germination.

Stage I (Leaf) Liquid

The formula we have developed for the first stage of growth, referred to as the leaf stage, supplies a sugary nitrogen source for increased soil bacterial activity and nutrient release along with some calcium and trace minerals. This gives plants a boost when they are lagging in growth force (which is a relatively rare condition on our farm). We often leave out the liquid fish, and sometimes we prepare a simplified blend of molasses, seawater, and milk only. Calcium nitrate or sodium nitrate could be substituted for the liquid fish. We apply this formula alongside the growing crops using the flat fan-spray-pattern hose ends, in a similar manner to the liquid seed starter. This volume of liquid will cover roughly 5,000 square feet (465 sq m) of growing crops.

80 fluid ounces (2.4 L) liquid fish (or other appropriate soluble nitrogen material)
160 fluid ounces (4.7 L) molasses
20 fluid ounces (600 ml) vermicompost extract

160 fluid ounces (4.7 L) seawater (or 10 ounces agricultural sea salt)
20 fluid ounces hydrated lime (600 ml) (caustic material, care in handling required)
160 fluid ounces (4.7 L) raw milk

Stage II (Flowering) Liquid

This liquid side-dress fertilizer helps bring a crop into flowering, or it can be applied once flowering has commenced. The milk and rock phosphate provide an increase in calcium and phosphorus. Along with the phosphorus, the micronutrients encourage flowering. The molasses serves as a high-carbon buffering agent. The rock phosphate is very finely ground. With occasional stirring, it stays suspended in the liquid to some extent. We make a similar liquid to promote fruit sizing (stage III), using the addition of ash from burning high-silica materials to provide a reactive potassium and silica source.

1 gallon (3.8 L) raw milk
1 gallon (3.8 L) molasses
8 fluid ounces (240 ml) liquid seaweed
1½ gallons (5.75 L) soft rock phosphates

8 fluid ounces (240 ml) soluble manganese sulfate
4 fluid ounces (120 ml) soluble zinc sulfate
2 fluid ounces (60 ml) magnesium oxide with 8 fluid ounces vinegar added to react

Stage I (Leaf) Solid Side-Dress

Due to our environmental and soil conditions, solid side-dressing materials are all compost-based. They are essentially compost piles specifically built for the various periods of growth. Often they are relatively small, contain many added minerals, and are assembled and utilized within short periods of time.

It's rare that we need to side-dress crops to promote leaf growth, but here is a formula we have used for that purpose. The compost for this recipe is one that included a high percentage of cattle manure in the initial mix. Feeder roots of crops reach into this material very quickly.

1 wheelbarrow full of manure-rich
   compost
1 quart (950 ml) molasses

2 quarts (1.9 L) coffee grounds
A splash of seaweed

We try to mix this material up a few days before use, but it can be mixed and applied immediately if necessary. Usually we make a pile on the ground and hand-mix with pitchforks, but if a particularly large batch is needed it can be assembled and mixed with a loader tractor. Scoop the mixture into buckets, rake back the mulch alongside the plants, apply the side-dressing, and then return the mulch to place. Though application rates vary, a 5-gallon bucketful is usually enough to side-dress about 40 row feet (12 m). It is best to water it in as well, as this continues biological activity and encourages crop roots to access the material.

Stage II (Flowering) Solid Side-Dress

This solid material for side-dressing during the flowering stage utilizes compost prepared for general field spreading along with added mineral materials. Because this material is very rich in mineral amendments, it is used at a lower rate than the Stage I Solid Side-Dress, often about 5 gallons (20 L) to 80 row feet (24 m). It is best to prepare this blend a few weeks ahead of time and let it sit. In a pinch, though, it can be mixed and applied immediately. We often use the loader tractor to assemble this material, and we apply it in the same way as the Stage I side-dressing.

1 wheelbarrow high-carbon
   compost (there are 3 wheelbarrow loads in one of our loader
   tractor scoops)
2 quarts (1.9 L) soft rock phosphate
6 quarts (5.7 L) gypsum
8 quarts (7.5 L) diatomaceous earth
1 quart (950 ml) dry seaweed

2 quarts (1.9 L) wood ash
8 fluid ounces (240 ml) molasses
4 fluid ounces (120 ml) sodium
   borate (Solubor)
2 quarts vinegar mixed with 6 fluid
   ounces (180 ml) magnesium
   oxide
8 fluid ounces manganese sulfate

Stage III (Fruiting) Solid Side-Dress

It can be logistically difficult to apply a solid side-dress material alongside plants in the fruit sizing period, but we have developed a recipe we can make use of when we can reach the plants. We use very old compost material for this blend, often materials that have been left behind from our more active composting piles. High-silica ash from the burning of materials from the exteriors of plants is preferred, but regular wood ash will serve.

1 wheelbarrow fully aged compost
1 quart (950 ml) bonemeal
1 quart (950 ml) soft rock phosphate
1 quart (950 ml) raw milk
2 quarts (1.9 L) dry seaweed
1 quart (950 ml) wood ash (high silica content if possible)

Foliar for Fruit Ripening and Disease Prevention

Here is an example of a foliar spray recipe that would be used for fruit ripening and disease resistance in August. Water for foliar treatments is probably best if it is not treated with antimicrobial materials like those found in municipal water. Well water is our choice, though active, healthy stream water may be even better. Rainwater in our area is so highly contaminated that it may well be antimicrobial. Rainwater in other regions may work well. If possible the well water is set out in the sun for a day before utilizing in the foliar sprays. This leaves the water in the best possible state for use in our conditions. About 15 to 20 gallons of foliar spray is applied to the acre (140–187 L per ha), so the appropriate number of 5-gallon buckets are filled with about 3 gallons (11.3 L) of water apiece, and materials are then added in proper proportion. All the following volumes of materials are for addition to 3 gallons of water.

1 pint (475 ml) seawater or 1 fluid ounce (30 ml) mineral sea salt (Sea-90)
4 fluid ounces (120 ml) high MgCl seawater (Sea-Crop)
1 fluid ounce (30 ml) liquid seaweed (highly concentrated)
1 fluid ounce (30 ml) apple cider vinegar
½ fluid ounce (15 ml) vinegar eggshell extract
½ fluid ounce (15 ml) vinegar bone extract
2 fluid ounces (60 ml) fulvic acid extract
½ fluid ounce (15 ml) liquid soap

½ fluid ounce (15 ml) Oriental Herbal Nutrient
1 fluid ounce (30 ml) FPJ (1 fluid ounce each comfrey, nettle, plantain, purslane, rutabaga)
1 quart (950 ml) horsetail tea
1 quart (950 ml) extracted IMO (indigenous microorganisms)
6 fluid ounces (180 ml) raw milk
2 fluid ounces (60 ml) raw honey
Dusting of Biodynamic 501 (ground quartz preparation)

For trace elements we utilize various ocean materials including seaweed and seawater. The seawater is collected from the coast of Rhode Island, which is about an hour's drive for us. The coastal area is relatively unpolluted, and thus the seawater gives a living fertilizer material. If we do not have seawater available, Sea-90 agricultural sea salt is added instead. Seawater contains many important trace elements that are useful for balanced crop growth and disease resistance. The biology of seawater is also potentially of benefit, so fresh living seawater is preferred over dried sea salt. Another ocean water material is Sea-Crop: seawater that has had its sodium chloride removed as well as a fair bit of its water, leaving a concentrated mineral salt relatively high in magnesium chloride. This form of magnesium is our favorite for foliar application as it is more readily available than other magnesium-containing materials. The magnesium promotes flowering and photosynthesis and the chloride helps with disease resistance, along with the other trace minerals in the concentrate. These materials also greatly assist in flavor development. Another ocean material is liquid seaweed concentrate, which is also full of trace minerals but has the additional benefit of containing the growth-enhancing hormones from the seaweed.

Once the ocean materials have been added to the water, we move on to the vinegar-containing materials. Vinegar promotes the flowering response in plants and is very useful for hardening plant leaves, which assists in disease resistance. However, too much hardening can make leafy vegetables less palatable for customers, so care is again in order. The formula lists 1 fluid ounce of apple cider vinegar, ½ fluid ounce of vinegar eggshell extract, and ½ fluid ounce of vinegar bone extract for a total of 2 fluid ounces of vinegar materials. This is flexible; anywhere from 1 to 3 fluid ounces of any mixture of these three materials may be appropriate. Less if leaf hardening is not desired, more if disease is threatening the leaves. Apple cider vinegar is used, but other vinegars may work just as well. Often we make the vinegar ourselves, though for the eggshell and bone extracts, commercial vinegars may extract the materials better. In order to make vinegar eggshell extract, we roast eggshells in the oven for 30 to 40 minutes or until they are a golden-brown color. Then they are crushed and added to 10 times their volume of vinegar. The acids react aggressively with the alkaline calcium in the eggshell. This forms calcium acetate, which is a highly available form of calcium and is useful for both fruit sizing and cell wall integrity. The mixture is allowed to sit and react for seven days, after which it is strained and stored for use. The vinegar bone extract is prepared in a similar manner. The bones are thoroughly charred in the woodstove firebox by filling a 1-gallon olive oil tin and placing it into the fire. Once the bones have charred, they are crushed and added to 10 times their volume of vinegar.

This material not only contains much calcium but also phosphorus, which is a critical element for proper fruit formation. Again the mixture is strained and stored after seven days. These extract recipes are originally from Korean Natural Farming techniques.

Liquid fulvic acid extract is a long carbon chain material that helps buffer the salts and vinegar solutions from damaging the biological-based materials that are later added to the spray solution. The fulvic acid also helps to keep the nutrients available for plant uptake. Fulvic acid extract is commercially available; if we do not have it on hand, however, we will utilize leachate from watering our vermicompost worm bins instead—about 8 fluid ounces (240 ml) per 3 gallons (11.3 L) water. This has a similar buffering capacity, though it may not be quite as effective at nutrient delivery. Along those lines, the liquid soap in the formula helps to keep the foliar spray on the leaves for absorption, and also helps prevent any oils in the mixture from coagulating. Once these ingredients have all been added a quick stir will help to prepare the mix for the biological ingredients to follow.

Oriental Herbal Nutrient, or OHN, is a fermented, extracted herbal preparation also from Korean Natural Farming. The basic recipe is to mix the herbs angelica, cinnamon, licorice, garlic, and ginger with fresh beer or sake, each in a separate container at a ratio of 4.4 pounds (2 kg) dried herb to 1.6 to 2.1 gallons (6–8 L) fresh beer or sake. (If using fresh garlic or ginger use more—about three times the dried weight.) Cover the containers for one to two days with a porous paper, then add a high-quality brown sugar at a ratio of 1:1 by weight and let sit another four to five days. At this time add 4 to 5 quarts (3.8–4.7 L) of vodka or other distilled alcohol, change out the paper covers to a plastic sheet secured with a rubber band, and let sit for another two weeks, stirring the mixture once a day. Then strain and store. (More thorough recipes for the Korean Natural Farming preparations are available either through Cho Han Kyu's manuals or on the internet.)

The OHN is utilized at ½ fluid ounce (15 ml), which is prepared by mixing all the separate herbal extracts to achieve this volume, with a little extra on the angelica. These herbal extracts are very useful for the vitality and immune system of the plants, and follow closely along the lines of herbal medicine.

Another plant extract from Korean Natural Farming is fermented plant juice, or FJP. Fresh plant material, preferably gathered in the early morning with dew still present, is chopped and mixed with organic brown sugar at 1 part plant matter to 2 parts sugar by weight. The separate plant materials are placed in a crock with a porous paper lid and allowed to ferment for seven days, at which time they are strained and stored. The sugar extracts the juices from the plant material; clues from herbal medicine provide the keys to which plants are best for extraction, along with a little intuition and guidance. Comfrey, nettle, and plantain leaves are extracted for general vigor and health, with rutabaga added here for fungal disease resistance and purslane for fruit coloring. One fluid ounce (30 ml) of each plant extract is added to the mixture.

From Biodynamics comes the addition of horsetail tea. This is 10 fluid ounces (300 ml) of dried horsetail herb simmered in 1 quart (950 ml) water for 20 minutes then strained and cooled. The horsetail is very high in an available silica, which provides a leaf hardening that greatly assists in resistance to fungal disease. Along these lines we also add a dusting of Biodynamic Preparation 501, a ground quartz material that also assists in the silica principle. Raw honey is added because it contains components of a plant-stimulating nature. Raw milk is added for the benefits of calcium and other growth nutrients as well as a powerful biology of beneficial nature. This brings us to the final ingredient: extract of indigenous microorganisms (IMO), another preparation from Korean Natural Farming. IMO is a culture of forest microbiology that is increased to a great degree through culturing steps and then extracted in a bucket through the addition of water and vigorous stirring followed by straining. It is similar in principle to a compost tea but contains a different microbiology that we prefer to compost teas. Preparation of IMO is fully described in chapter 11.

Now that all the ingredients have been added to the water, a vigorous stirring of the mixture is applied in a Biodynamic fashion. That is, a wooden stick is used to swirl the mixture in one direction to create a vortex; then after a vortex is established, the direction is reversed, causing a thorough mixing of the ingredients and an introduction of air. With this stirring and mixing comes the opportunity to impart the force of will or prayer into the material, so this is a time for concentration or maybe a song. The mixture is stirred upon assembly for about 5 to 10 minutes and then applied to the fields within about an hour. The oxygenation and timely spraying help the aerobic organisms from the IMO survive. Usually four or more people are involved in the mixing and spraying of 3 acres (1.2 ha).

FIGURE 9.3. Creating the vortex. Time for concentration and prayer. Working barefoot can be a great advantage in lightness of foot and general maneuverability in close-spaced crops; it also allows water, nutrients, and earth energies to flow into the individual.

The spraying occurs in the morning as early as possible—well, at least by around eight o'clock—as the plants are receptive to foliar materials at this time. Evening can also work but we generally do not want to wet the leaves further at that time. Backpack pump sprayers from Jacto are used to apply the foliar. These are quiet and allow for a peaceful concentration as the person spraying again has the ability to impart prayers into the material as it wafts through the air onto the leaf's surface. The foliar spray is a blessing for the crop. The spray wand is moved back and forth in an arc that covers about 3 feet of bed surface on either side of the sprayer. Jacto sprayers have a small agitation device that helps keep the foliar mixed in the tank, and we have adapted them by putting a splitter on the wand, giving us two tips. The sprayer tips have larger openings than the ones supplied originally as well. These parts are available from Jacto. This allows for more spray to be applied per pump, yet still gives an appropriate mist. The foliar mixture is strained through a fine paint strainer mesh; material that settles to the bottom of the buckets is not poured into the sprayer. However, sometimes the mesh filter in the sprayer tips does clog and needs to be cleaned out. This is performed in the field by using the foliar liquid as a wash from the disassembled wand head. Upon completion the sprayers are washed out and clean water is pumped through the spray wands.

Foliar applications start in the spring with application every two to three weeks, then intensify to about weekly during the August period of fungal disease, and usually end about the time of frost when covers are being applied over the vegetables. When the low tunnel covers are down for a rain or other event in winter, a foliar spray may be applied then as well. Though a relatively complex recipe was provided here for an example, simpler formulas may certainly be utilized depending upon conditions—say, seaweed and milk or seawater and vinegar. Again, much adjustment is possible to suit individual conditions.

In terms of foliar applications, Biodynamic Preparations 500 (horn manure) and 501 are very useful for assisting in the balancing of the field conditions. They are powerful energetic materials. Preparation 500 greatly enhances the growth force polarity and is most frequently applied as a liquid before seeding or otherwise early in the growth period of the crops. Preparation 501 (horn silica) greatly enhances reproductive polarity and is foliarly applied to crops at flowering. We apply both of these materials at appropriate times by themselves as well as blending them into other formulations. The preparations can be made on the farm or purchased from Biodynamic suppliers. The Biodynamic literature is full of preparations, methods for making them, and in-depth application principles, which are well worth reading. However, the basics are to stir the appropriate volume of the preparation in water for one hour in a

FIGURE 9.4. Foliar material is applied in a sweeping arc, back and forth. A blessing for the crop.

back-and-forth vortex action with intent and focus. The 500 is applied in the evening, with a bucket and whisk broom for large droplets. The 501 is applied in the early morning after stirring, using the backpack sprayers, in the same manner we apply the foliar sprays. These Biodynamic spray preparations are very helpful in their ability to help reestablish balance.

Composting

Composting is of course the very heart, backbone, and, shall we say, digestive tract of the organic system of agriculture. Most soils benefit to a great extent from the application of high-quality farm-made compost. The benefits of the biologically stimulating nature of compost are quite significant in terms of yield and quality, especially when combined with materials that aid in the mineral nutrient balance of the soil. Compost can indeed imbalance a soil if the materials utilized are in severe imbalance, so guidance here may be appropriate. However, often the best way to learn these things is by doing: Make the compost, trial it, evaluate and learn from error and success.

The compost pile is the stomach of the farm organism. It is where organic materials are mixed and decomposition begins, to be followed by nutrient absorption in the field. The application of compost along with mulch materials and crop residues is the primary means of keeping the soil life fed and productive. Thus enhanced, a field's biological function provides effective balanced nutrient release for healthful crop growth. The results of compost application can be enormous: dramatically improved biological function, soil structure improvement and a resulting enhancement in soil air and water conditions, additional nutrients released for crop growth, and better soil temperature regulation, just for starters. The composting system also provides a place well suited for the introduction of mineral materials to be assimilated into the living systems. Although too much of a good thing is not of benefit, having too much compost is a rare condition, and there is great need for more compost in today's agricultural systems.

Purchasing commercially produced compost may be a worthwhile investment, particularly when growers are just getting started and soil quality is poor. Caution is in order, though, because purchased composts are variable in quality. The materials may be too young or too old or be made from inappropriate materials that do not achieve appropriate balanced growth. They often contain objectionable materials such as plastic garbage (which needs to be picked out) or contaminants (which may be difficult to determine the origin of). It's advisable to trial a limited amount of a commercial compost to observe its impact on crops and soils before making a substantial purchase. Composts can be expensive to purchase, especially well-made ones, and a new vegetable field with poor soil may benefit from an application of up to 100 tons per acre (247 metric tons per ha).

The growing of vegetables invariably produces excessive bulky materials, and all vegetable growers will thus encounter the need to develop their own composting system to deal with these materials as well as other farm or household waste or byproducts. Indeed, considering costs, quality and nutrient balance issues, and the need to deal with on-farm wastes, it is very useful and beneficial for growers to take composting into their own hands.

Compost Management Basics

In the composting process organic (carbon-containing) materials are piled up in appropriate mixtures to achieve effective heating and associated increased speed of decomposition. It is up to the composter to provide the right conditions and monitor progress. Tasks include the balancing of appropriate materials (often referred to as balancing the carbon to nitrogen ratio), maintaining the moisture level of the pile, and providing porosity or air space in the pile. The carbon to nitrogen ratio of a pile is a way to express the need for a blend of appropriate materials. The ratio is often placed around 30:1 carbon to nitrogen. This ratio gives the grower a guide to volumes and is not usually analyzed by a laboratory for most composting systems. The indication is to balance nitrogen-rich materials with a generous amount of carbon-rich materials. It's a pretty rough guide, and successful composting happens as long as the balance is somewhat close to that range. If the C:N ratio of a pile is too low, there will be foul odors, likely an ammonia (NH_3) smell. If the C:N ratio is too high, the pile will be prone to drying out and will decompose slowly. When the C:N ratio of materials is in the target range, temperatures in the pile will rise significantly. This heating is what distinguishes composting from ordinary decomposition of mulch, sheet compost, and crop residues that happens at the soil surface at field temperatures.

The compost pile is an excellent place to add minerals that may be needed to help balance the nutritive condition of the soil where the compost will be applied. The vast majority of mineral fertilizer additions to our field happen through compost. This allows for digestion of these minerals into more accessible and biological forms and is very useful for the more difficult-to-access minerals like the silicates. It is also very useful for buffering the potential damage that soluble salt minerals could inflict upon the soil biology if applied directly. If a compost pile is planned for broadfield application, then the mineral materials to achieve general field balance can be added. Therefore different compost piles receive different mineral additions depending upon place and time of use. If the compost is for side-dressing, then particular mineral materials are added for that purpose. The potting soil compost receives its specific minerals, and the vermicompost its specific minerals. It is a skill to build an appropriately balanced compost pile for specific conditions. The foundational presentations of concepts in of earlier chapters has hopefully provided the framework to help guide decisions regarding the choice of materials to blend in a compost pile.

MONITORING TEMPERATURE

The intense heating of compost is the result of accelerated microbiological activity in the protected pile environment. Heat from decomposition does not readily dissipate in this environment, which results in more microbe activity and even more heating in a feedback loop until stabilization is reached. This often can be as high as 160°F (71°C). The factors that influence pile temperature include pile size, components, moisture level, aeration, and porosity.

The size of the pile impacts heating by providing a protective environment for heat to amass, sheltered from cold ambient air temperatures and winds. If a pile is too large, however, then air will have difficulty reaching the center of the pile, which will inhibit decomposition at the center. The optimal size of the pile is thus dependent upon the structure and porosity of the materials as well the accessible nutrient levels. In general our piles on the farm are about 6 feet (2 m) tall and 12 feet (4 m) wide at the base, arranged in rows of varying length. With lots of porosity of materials, this generally gives us consistent compost decomposition into the heart of the pile.

Pile composition affects heating in more than one way. Some carbonaceous materials, like big wood chips from tree heartwood or sawdust from kiln-dried lumber, have very high carbon content, but it is not readily accessible by microbes, and this may reduce heating. The composition of the materials is also the primary factor in pile porosity, which in turn has great influence over the air:water balance and temperature. Coarse-textured materials such as straw and wood chips increase porosity, but materials that compost readily, like manure and even matted wet leaves, may diminish porosity. If porosity is diminished too much, pile temperature decreases.

When there is too much air in a pile (usually due to too much carbon), molds often form; decomposition is generally slow and weak, and temperatures may be cool. When a pile is too wet, the result is very similar to too much nitrogen. Temperature is lower than optimum, and putrefaction occurs with foul smells. Growers can shift the air:water ratio of a pile by turning it, which tends to aerate and dry the pile overall. Thus if a pile is too wet growers can turn it to assist with drying, which could lead to an increase in temperature. Additional coarse materials could also be added. Exposure to precipitation has a dramatic impact on the air:water ratio. Growers can irrigate piles if they become too dry or cover the piles to keep in or keep out moisture, depending on the objective.

Piles usually start to heat within a day or two after assembly and will spike up to a stable high temperature for a period of time. The temperature will then steadily drop, leveling out close to ambient temperature. This temperature increase and decline can take anywhere from weeks to months. If a quicker cycle is the goal, turning the piles should help, but for most growers a slower, less labor-intense process lasting several months is often the method of choice.

SETTING UP A COMPOSTING SITE

To produce compost well in quantity, it is necessary to set up an appropriate composting area or yard. The compost yard needs to be large enough to stockpile all the ingredients in separate piles and also allow sufficient space for building the mixed piles, moving them during the turning process, and stockpiling mature compost for later usage.

FIGURE 10.1. The compost yard has a solid base, no weeds, and many piles of the various materials.

If large amounts of materials will be hauled in from off-site, the yard should allow easy access for trucks. This is very important, because truckers absolutely hate to get stuck! Not only that but easy access with a secure base makes truckers happy, and happy truckers fill their trucks fuller, which means more compost!

Compost piles do well when placed directly on top of soil. This allows for biological interaction of soil organisms with the pile, and helps to maintain pile moisture. Beneficial biology develops on the site, including diverse organisms that play a role in all the different stages of decomposition; this helps with re-inoculating succeeding compost piles. For small-scale composting it may be possible to build piles in different locations where the exudates and residues of the pile would be of benefit to the soil under and around the pile. Larger-scale composting yards are often built with a base of stones, such as cobblestones from the fields, with gravel on top of the stones. This type of base supports heavy tractor and vehicle traffic while still having a soil-like interface that allows some biological interactions—which a concrete pad, for example, does not. The gravel surface also helps minimize weed growth around the piles. Some gravel may be scooped up when piles are mixed or turned, but if the operator is careful, the volume of gravel incorporated into the compost will be minor. Actually it's potentially a benefit, because it adds a little more diversity to the compost mixture.

Sourcing Materials

The recycling of once-living materials from the local environment is at the heart of the organic system of agriculture. Though conditions have become increasing imbalanced and local materials may not support growth the way they once did, it is still of great benefit to engage in the traditional agricultural pursuit of recycling local organic (carbon-containing) materials as the basis for fertility. In today's environment, contamination of materials with detrimental substances is unfortunately common. Growers need to determine whether the level of contaminants are beyond the ability of the compost and soils to detoxify. The best way to determine this is to learn as much as possible about the source of the raw materials. Materials can be physically examined. Another way to assess their effects is to set up a trial, applying the raw materials around growing plants and carefully observing the impact on growth. Chemical contamination often leads to distortion of growth. The finished compost can also be biologically assessed in this manner.

The intense decomposition that happens in a compost pile is effective at breaking down many contaminants, and a highly functioning field soil also has this ability, as well as the gift of being able to help keep toxins from cycling into the process of life by bonding them to clay-humic substances. It is best, however, not to challenge the soil and compost systems in this regard if possible.

The high-carbon or "brown" materials utilized in composting make up by far the bulk of a pile. These materials are usually fluffier than the nitrogen-rich materials and are used at a high percentage to keep up a pile's C:N ratio as well as the porosity. The most common compost materials of a brown nature are straw, leaves, wood chips, shavings, and sawdust. The common nitrogen-rich materials utilized in on-farm composts are manures, vegetable and animal residues, and grass or hay. They are often referred to as the "green" or slop portion of the compost pile.

Stockpiles of nitrogen-rich materials generally require more careful attention while they await compost assembly. The nitrogen-rich materials tend to become anaerobic and putrefy if not balanced with carbon in a timely manner. This results in odors that are quite pungent, often odors of ammonia and sulfur, which represent nutrient potential lost as well as possibly offending the neighborhood. Stockpiles of nitrogen-rich material are also prone to leaching under heavy rainfall, which is yet more nitrogen loss and has the potential to disturb soil where the excessive runoff infiltrates. Thus it is of benefit to mix some carbon materials into nitrogen-rich materials immediately upon delivery. We lay down a carbon base, usually wood chips, and pile the materials with leaves or other carbon material mixed in as well. This begins or otherwise continues the materials heating and essentially pre-composts the materials before assembly. Temperatures of 150°F (66°C) or more are common in these piles. This stabilization and pre-composting of the initial accumulated materials should not continue for too long or the nutrient release and ability to heat the fully assembled compost piles will be diminished. Some growers utilize rock powder such as gypsum or rock phosphate to help stabilize nitrogen-rich materials at this stage. We hold off on adding these materials until full compost pile assembly. Carbon stabilization, as well as covering piles of N-rich materials with carbon-rich materials, also greatly aids in odor and fly control.

STRAW

Straw is the stems of annual grasses grown for grain. After harvesting the grain, some grain growers will then bale the leftover crop stems. This often results in grain seed mixed into the straw. There is also a potential for weed seeds if growing conditions were not weed-free. This is generally not a problem if there are only a few weed seeds because composting breaks down most (not all) seed during the composting process, but is something to watch. Rye, wheat, barley, and oat are the most common straws, but corn silage is essentially a straw as well and can be used in compost making, though it is slower to decompose than other straws. Allowing the straw materials to be exposed to rain and the environment and thus pre-compost may be useful in speeding their decomposition

168

in assembled compost piles. Otherwise they can be mixed into piles from dry storage. We often use straw as the carbonaceous "skin" cover of compost piles after initial assembly. Straw is well suited for this because small bales can be brought upon the top of the pile and then shaken out over it with relative ease. The straw covering then serves to shed some rain if excessive, similar to a thatched roof. At the same time straw can retain some moisture in the pile and is permeable, allowing air exchange.

Straw is often contaminated with agricultural chemicals. Pesticides in particular offer difficulties here, especially the herbicides. As explained in chapter 6, many herbicides disrupt biological and chemical function in the soil and result in straws of imbalanced nutrient profile, but far worse for vegetable growers is the ability of some herbicides to resist decomposition in the compost pile and thus give an herbicide burn to vegetable crops. A particular group of herbicides is known as the *persistent herbicides*. These same herbicides can contaminate hay, grass, and manures from animals fed on the grains grown with them. Herbicide burn may be very difficult for growers to diagnose as often it is at a relatively low level, but it is still yield reducing and of serious concern. Growers need to be aware of the potential for these chemicals to be in their straws and other composted materials.

One approach to being more certain of straw quality is to grow it yourself or have another trusty grower produce it for you. In this case grain is not the pursuit and the straw can be cut before reaching maturity, which yields a straw free of grain seed. Many farm operations have the necessary equipment to produce this straw because no combine is needed, just the standard hay cutting and baling equipment. In this relationship quality can be best assured. One grower we have been working with not only controls weeds without herbicides but is also providing balanced nutrients for abundant straw crops with an appropriate nutrient profile for our fields. On top of this he double-crops some straw fields, following winter rye with a summer crop of sudangrass, thus yielding much straw per acre per year. Vegetable growers can grow their own straw if they have enough land and the appropriate equipment, but in many situations vegetable-growing land is too valuable to plant with straw crops, and vegetable growers too busy tending high-value crops to engage in this pursuit. On our farm we do produce a high-quality straw from some of the overwinter cover crops. This is primarily winter rye with crimson clover and vetch mown in May before June-seeded vegetables. However, we chop this material in the field and use it for mulch; little is left over for use as a compost ingredient. Straw can be purchased from farms in grain-growing regions, often at reasonable prices, and in large bales, but trucking expenses may be high. Of all the compost materials we utilize, straw is the most expensive, and thus we use less of it than leaves or wood chips. We add it to provide some beneficial diversity in our mixture.

LEAVES AND WOODY MATERIAL

Tree leaves are an excellent compost addition. Leaves are nature's primary soil fertilizer in forested regions, and soils developed in these regions are likely to benefit from relatively profuse use of leaves in composts and mulches. Leaves are frequently highly abundant, often free for the taking, or for a small hauling fee. The main contaminant in leaves is plastic garbage and other debris that will not decompose. Leaves from roadside collection in urban environments are more prone to this type of contamination. The garbage can be picked out or possibly screened out. Forging a personal relationship with a landscaper can help secure a quality supply of leaves with fewer contaminants. Leaves themselves offer little direct chemical contamination but leaves mixed with grass may be contaminated with herbicides. Leaves are commonly collected in autumn, so ample space needs to be provided to pile them up leaves at your compost site, especially if you have a supply of landscaper-collected leaves. This avoids any unnecessary double handling.

The relative coarseness of chopped leaves impacts their usefulness on the farm. Most landscapers and municipalities collect leaves using large vacuum machines that partially grind them. This is absolutely excellent for composting, because unchopped leaves may mat excessively in compost piles. Mowing machines with a bagger component are also used for collecting leaves, and in this case they are more finely chopped. Sometimes mown leaf is dumped in piles and then vacuumed up for a double grind! This well-ground material is ideal for field mulching, so we set such loads aside for that purpose instead of using them for compost making. Leaves can be run though grinding equipment at the compost site, but this is generally not efficient. We have run whole leaves through our bale shredders, but watch out for small stones! We have let whole leaves sit in a pile and partially decompose before use; this creates material similar to ground leaves. Sometimes leaves from municipal stockpiles are also partially decomposed or otherwise available in various states of decomposition.

The trunks and branches of trees are another common component of composts in forested regions in the form of sawdust, bulk wood shavings, and wood chips. Sawdusts are highly variable in their makeup. Softwood sawdusts decompose more slowly than hardwood, and both decompose even more slowly when they are a byproduct of kiln-dried lumber. Sawdust from cabinetmakers and other shops are more prone to contaminants such as glues; sawdust from the milling of saw logs is more likely to be free of contaminants. Sawdust must be carefully mixed into a compost pile because they have a tendency to stick together and form large masses that resist decomposition.

Sawmills debark logs before running them through the saws, and sometimes they amass excess bark. In our region bark mulch is so much in demand in the landscape trade that little is available for composting. Wood shavings are logs and wood materials that have been specifically ground or shaven in order to produce a material appropriate for bedding animals; they are commonly encountered mixed with manures. Wood shavings are more prone to compacting than wood chips and may not provide as much porosity. Be aware of possible contamination due to the grinding of wood "wastes" for shaving production. In our region this includes old pallets. Most wood shavings are still produced from clean materials, but it is worthwhile to closely examine shavings for strange coloration and smell. Or simply make sure you know your supplier and what the origin of the shavings is.

Wood chips are probably the cleanest of materials commonly available for composting. Forested areas receive little chemical treatment, trees are generally thought to take up less toxic materials than other plants, and there is usually little in the way of plastic garbage in wood chips. Chip size and makeup are variable and depend on what type of woody material is being ground, as well as on the efficiency of the grinder. A common source of wood chips is roadside clearing projects. These chips often consist primarily of the smaller twigs and branches of trees, which are much more nutrient-rich than the heartwood of older trees. This means that such chips decompose much more rapidly—all the more so if the material was ground in the summer and green leaves were mixed in as well. Other wood chips come from land-clearing operations. In this case the material has a much higher heartwood content and is thus slower to decompose. The chips of softwoods may decompose more slowly than many hardwoods, so it is important to ask what species of tree is predominantly being chipped. Cedar and walnut may be very slow to degrade and may not be appropriate in large volumes. Wood chips can be pre-composted or aged before being incorporated into the compost system; this will speed their decomposition. As with any compost ingredient, too much aging will diminish their value as food for the hungry compost microbes and soil organisms. Most wood chips can sit for a year or two without losing much value as a compost ingredient.

MANURES

Manures are a traditional nitrogen source for composting, and the benefits of manure in composting and overall farm fertility can be dramatic. Many livestock producers are happy to give away manure, but the hauling cost often needs to be paid. If a manure is particularly high-quality or otherwise the most appropriate for a grower's composting system, it is usually worthwhile to pay the cost of hauling and even a fee for the manure itself. The benefit of manure is one reason that growers might choose to integrate livestock into their system, as has traditionally been practiced.

Manures contain an animal-selected, partially decomposed feed source for the compost pile as well as beneficial microbiology for both composts and field. Much of this microbial richness comes from the animal's digestive tract. Cattle, cows, and other ruminants, with their extensive digestive systems, are most renowned for their microbial diversity. As tremendously beneficial as manures are, growers need to be aware of appropriate balancing of nutrients and contamination issues. Contaminants include antibiotics, anthelmintics (deworming drugs), insecticides for fly control, and hormones, as well as persistent herbicide residues in the feeds. All of these are detrimental to vegetable crops and soils. Although the biological activity in composting and healthy soils is able to decontaminate these materials to some degree, it is best to not challenge our systems in this way. Manures require particular attention in this regard. Many livestock are given feeds that may be unbalanced in terms of nutrients, which creates concern for the nutrient balance of their manure and its potential influence on growing conditions. Some feeds contain additives of particular toxicity, such as the arsenic supplementation of chicken feed. Again, building personal relationships with livestock producers and knowledge of their techniques is in order to secure high-quality manure.

Cattle manure is considered the easiest manure to compost, and it does indeed make composting very easy, exemplified by its ability to quickly heat up. It will start almost any pile cooking. Historically cattle manure has been the manure of choice for vegetable growing in our region of Connecticut, and many others. Spent hay is often found mixed in with cattle manures, which is of benefit to help the manure begin heating, but may contain weed seeds as well. Grass-fed cattle manure is often the best in our region. These animals are generally quite healthy and receive minimal veterinary care. Thus there are fewer contamination concerns, and the manure is a well-balanced source of nutrients with excellent microbiology. Some grass-fed cattle are pastured year-round and often eat the farm's own grass hay during the winter months. There is usually a winter feeding area for the hay where manure-hay mixtures pile up and can be collected. We source manure for our composting operation from a local producer who has installed concrete feed pads in the pastures with a concrete block barrier to push the manures against for loading. The farm has well-developed farm roads, too!

Cow manures are similar to cattle manures, but the manner in which dairy animals are kept often impacts the content of their manure. Although there are a few grass-fed dairies around, more commonly cows are generally fed a much more intensive diet to maximize milk production. The feed often contains grains, which impacts the animal's digestive microbes and changes the mineral profile of the manure, raising the phosphorus level in particular. Again, the feed source needs to be carefully considered for its balance and potential impact on the vegetable-growing soil. Veterinary chemicals of potential concern are much more common in dairy animals than with beef cattle. Manure management on dairy farms is variable. Sometimes the solid manures are almost pure in their nature but other management methods result in various amounts of bedding materials soaked with urine mixed in with the solids. Government regulations have mandated that many dairies install manure lagoons. Moving manure into lagoons is very detrimental to its quality, and it is best to avoid manure from lagoons if possible.

Horse manure is not traditionally highly regarded as a component of composts for vegetable crops. It has low nutrient value and is prone to excessively high heating, which alters the biological development of the compost pile. (Horse manure has been traditionally used in hotbeds for growing vegetable crops in the colder months.) Veterinary care and thus use of drugs is more common with horses than other livestock. Many horses are on dewormers constantly, for example. Horses are regarded as finicky eaters, with the result that "horse hays" are often grown with more chemical treatment in the form of herbicides. There are often large amounts of bedding mixed into the manure as well. Horse manure is highly available—around our area some people are paid to haul it away! Some growers may still choose to try to work with it, but it is best to know its qualities. Horse manure also is the main ingredient of many commercial "mushroom composts" from commercial mushroom growing.

Poultry manures are by far the most nutrient-rich of the commonly available manures. Commercial poultry is fed a diet of almost entirely grain, with other concentrates added. Therefore the manures are very high in nitrogen and phosphorus, just for starters. Laying hens are often fed calcium materials of a mineral nature as well, making their manures also high in calcium. It is interesting to note that chickens' digestive systems—particularly the grinding action of their gizzards—prepares them to digest mineral materials better than other animals. Obviously chicken manure from factory-like production is generally contaminated with chemicals. As mentioned, arsenic has long been fed to chickens, simply because it makes them grow a little faster. Some chicken factories feed insecticides to the chickens just to kill flies in the manure. Chickens raised in factory conditions do generate a lot of manure, however, and if a grower can secure higher-quality manure from chicken producers, it may be well worth working with. There is little to no bedding material in poultry manure, so it needs to be mixed with plenty of carbon material to break it up and dilute its high nitrogen content. Although chicken manure is lacking in the beneficial biology of the ruminant that makes composting so easy, it can do well in compost piles of mixed manures. Poultry manure piles are very prone to breeding multitudes of flies, so the sooner it heats up and begins composting, the better. At Tobacco Road Farm the poultry forage on the compost piles, thus adding their own manures to the composting system.

Sheep and goat manures are sometimes available, particularly from dairies. These manures, being from ruminants, are similar in many ways to cow

quarries, and many kinds of minerals in the various rocks, but quarry workers usually know the basic makeup of the rock they are processing. In our region large volumes of high-silica rock such as granite and trap rock or basalt are processed. This material is particularly appropriate for us because the addition of biologically available silica into our fields brings greater balance, plant strength, and insect/disease resistance. Quarry dusts usually contain a large portion of clay and silt. Often the finest dusts are the least valuable byproduct of rock crushing and are quite affordable (three to five dollars a ton plus trucking in our region). Some rock dusts are collected from dredging of quarry ponds where material has accumulated during the washing of crushed stone. Quarry dust may be very clay-rich and may compact into hard clods, which may be difficult to separate for thorough mixing into compost. Some quarries' processes result in mixtures of coarser materials, and these may be easier to incorporate. Other rock minerals used as compost amendments such as rock phosphate or talc may also be ground finely enough to serve as clay. In addition, we backhaul clay leftovers from a potter in town. This dense clay material is called slip. It needs breaking up or liquefying by vigorously stirring in water before being added to the compost pile.

Assembling a Pile

When it is time to build a new compost pile, we combine materials in somewhat established proportions—a compost recipe, if you will—and then evaluate it and alter it if needed. Developing an appropriate recipe requires consideration of several factors. The carbon to nitrogen ratio will have to be considered in terms of both achieving active composting as well as the impact of the compost on the C:N balance of the area of the field where is the compost is intended to be utilized. This is also the case with the mineral additions. Another important consideration is how the compost will be used: top-dressed or incorporated into soil A compost that is not going to be incorporated into the soil allows for less decomposed material upon application. Top-dressed compost can be bulked up with the fluffy, slow-to-decay wood chips to allow for better air penetration into the pile. This reduces the amount of turning required and is an aid to the growth of fungal organisms. We also consider the fungal to bacterial ratio, which is highly associated with the C:N ratio. Fungal-dominated composts generally require a more carbonaceous mixture and lower composting temperatures. If the compost is meant for soil incorporation, as in a tillage system, or for use in potting soil, then hard-to-digest materials such as wood chips may be less appropriate, or more turning and/or lengthier maturation of the compost may be required. It can be quite detrimental to incorporate undecomposed materials—especially highly carbonaceous material—into soil,

TOBACCO ROAD FARM COMPOST

Here are the details of the compost we make at Tobacco Road Farm for broadcast, preplant application upon the no-till growing area. Note that some of these materials are salts added in very small quantities, but these small additions can significantly improve crop growth. (Take note that some mineral salts are available in both nonsoluble grade and soluble grade. If the intention is to liquefy the material for application, purchase soluble grade.) The formula is based upon years of field observation, tissue and soil laboratory testing, crop response, and spiritual guidance. This recipe is unique to our field conditions; it is not likely that this precise formula will be right for your soils and environment. It is provided here as an example of materials and amounts that have proven useful for our situation and to describe how we make and use this particular type of compost.

BASE INITIAL INGREDIENTS

30% wood chips
20% leaves and/or straw
40% cattle manure, pre-composted
    mixture
10% vegetable scraps, pre-composted
    mixture

MINERAL ADDITIONS
AT ASSEMBLY

(Add the following to about 30-plus cubic yards / 23 cu m of base ingredients.)
5–10% quarry dust and/or clay subsoil
    (1.5–3 cubic yards / 1.1–2.3 cu m)
100 pounds (45 kg) gypsum (calcium
    sulfate)
150 pounds (68 kg) calcium silicate
    (wollastonite)
25 pounds (11 kg) hydrated lime
    (calcium hydroxide)
250 pounds (113 kg) talc (magnesium
    silicate)
100–200 pounds (45–90 kg) soft rock
    phosphate

MINERAL ADDITIONS
AT TURNING
(1–3 MONTHS LATER)

250 pounds talc (magnesium silicate)
100–200 pounds (45–90 kg) soft rock
    phosphate
10+ pounds (4.5+ kg) elemental sulfur
50 pounds (23 kg) agricultural sea salt
    (Sea-90)
40 pounds (18 kg) manganese sulfate
    (soluble grade)
Up to 5 pounds (2.3 kg) sodium
    molybdate
5 pounds (2.3 kg) zinc sulfate (soluble
    grade)
5 pounds (2.3 kg) copper sulfate
    (soluble grade)
1 pound (450 g) sodium borate
2 ounces (62 g) cobalt sulfate
    (soluble grade)
1 splash of selenium feed supplement

because it alters the biological balance, resulting in the tying up of nutrients while the material decomposes. These tied-up nutrients may be required for crop growth. Thus, if undecomposed materials are worked into soils, a waiting period is probably in order before planting a crop in that soil.

It's best to accumulate all materials before assembling the pile, but if an ingredient is short or lacking altogether, we just substitute a similar material, such as leaves in place of wood chips or vegetable scraps for manure. We use a tractor loader, combining 3 buckets wood chips to 2 buckets leaves to 4 buckets cattle manure and 1 bucket vegetable scraps. The addition of the quarry dust or subsoil and other minerals and clays adds about another 10 percent volume, for a total of 110 percent, which is what most of our efforts add up to. Because the initial premix piles of manure and vegetable scraps also contain some carbon-rich materials, the assembled pile is quite high in carbon and moderate in nitrogen overall. The use of more carbon materials results in a generally lower composting temperature of about 120°F (40°C), which favors the more fungal-rich compost we're after due to our soil conditions. The initial premixed piles of cattle manure and vegetable scraps often heat to much higher temperatures (around 150°F / 65°C). When this hot material is mixed in with the rest of the compost ingredients, the temperature usually stabilizes around 120°F. The high volume of wood chips provides a base layer on the ground and a bulking of the pile that allows for adequate air exchange, thus greatly reducing the need for turning, which also is of benefit for fungal organisms.

We build up the piles into windrows about 12 feet (3.5 m) wide by 6 feet (1.75 m) tall with varying lengths. When beginning pile construction we first lay a base of wood chips about a foot (30 cm) thick. As in the example above, it is of benefit to have a carbon base material about 1 foot thick at initial assembly; this base serves to soak up nutrients that may leach down through the pile. We then pile the rest of the materials atop this, mixing materials with the tractor bucket as much as is quickly and efficiently possible. As the tractor operator is assembling a pile, two other people work to liquefy some of the mineral talc, hydrated lime, and wollastonite, and spray it onto the pile. This liquefying and spraying requires two people because some of the minerals do not go into solution in water, and they must be more or less constantly agitated by hand in a large stock tank. As one person stirs, a heavy-duty sump pump moves the slurry to the second person, who is working the hose to spray down the pile. This gives us very effective mineral distribution into the pile and greatly cuts down on our exposure to the lung-aggravating silica dusts and hydrated lime. We often apply gypsum and rock phosphate to a pile dry because they are less dusty, and this helps cut down the amount of minerals that need to be agitated. Usually either the person spraying or the person mixing can, in addition to their task, apply dry minerals as the pile is being assembled.

FIGURE 10.2. One pile is being turned and incorporated into another.

FIGURE 10.3. As piles are being assembled, we apply the liquid amendment between tractor loads.

The pile is left sitting for a month or more, and then we turn it, which is an opportunity to further mix the materials and also to apply additional materials as noted in the recipe and as we determine is needed. This is usually the only turning the pile receives. The piles are turned using a loader tractor, and basically the windrows are reassembled alongside their original position. As we turn the pile, we add more rock phosphate, and elemental sulfur as solids. We also mix a second tankful of talc and mineral salts and spray that on the pile. The salts include sea salt and manganese, zinc, copper, and cobalt sulfates, along with sodium molybdate and borate. Some of these salts need to be handled carefully: Don't get them in your eyes. Dissolving them and spraying them onto the pile gives an excellent distribution of these rather small amounts of material.

Introducing materials, especially the soluble salts, later in the compost process seems best because it allows the process to be well under way, and that minimizes the risk that the minerals will reduce biological activity. Also, the piles are better prepared to buffer and hold these soluble nutrients at this time in the composting cycle.

We monitor pile temperature by the use of 20-inch (51 cm) long compost thermometers. If the temperature climbs too high or fails to climb high enough, we adjust materials or techniques appropriately. The pile temperature stays relatively steady for a few months, often assisted by the one turning. The compost is ready for use when the temperature has dropped down below 90°F (32°C) or close to ambient, and the nitrogen-rich manure and vegetable scrap is no longer evident. Partially decomposed carbon material may still be visible; this is ideal for surface application, as these less decomposed materials provide excellent food for in-the-field soil biology. Our "finished" compost resembles a mulch/compost mixture.

Other conditions to monitor as the piles progress are moisture and air. These two factors are in relation to each other: When there is more air there is less water, and vice versa. Generally in our environment in the Northeast there is sufficient rain to suffice for composting. We rarely need to add moisture. Instead, we must cover piles during periods of excessive rain, which is often in the cooler months. We use large black plastic tarps, often silage tarps that originally served for covering straw or other materials. These used tarps invariably have a few holes, which are helpful in allowing the piles to breathe, but the tarps still shed most of the rain.

Air penetration into the pile is generally sufficient due to the coarse nature of the materials we compost, along with appropriate moisture control and proper sizing of the piles. Insufficient air will lead to anaerobic conditions, which result in off-smelling compost with a black color similar to swamp muck. If this occurs we recycle this material by adding it to a new compost pile in addition to the basic recipe. We also monitor the pH of the compost piles. This recipe usually

FIGURE 10.4. Compost adequately cooled, ready for usage, with many fragments undecomposed.

produces slightly acidic compost around 6.5. To make compost that is more acidic, one option is to effectually reduce the proportion of minerals (some of which are alkalizing substances) or by increasing the pile size (30 cubic yards / 23 cu m) while keeping the quantities of the minerals unchanged. Another option is to increase the amount of elemental sulfur. Doing the opposite would result in more alkalinity.

Achieving a homogeneous mixture of the diverse ingredients is the goal. This is relatively easy to do if compost is assembled by hand. However, loader tractors are almost always required in the making of large volumes of compost, as the lifting of tons of material by hand and back is laborious, though it could be thought of as healthful exercise—maybe. Manure spreaders can be loaded with material and then operated in a slowly forward movement so as to build windrows, and they do a fine job of mixing as they expel materials. However, we have not found them to be necessary, and they are slower than mixing solely with the loader tractor. There are also specialty compost turning machines utilized in large-scale composting where much turning is desired to produce compost in very short periods of time. Desired outcome often determines the equipment to be utilized. If fungal-rich, not fully decomposed compost for top-dressing is required, a loader tractor or even hand assembly may be sufficient. When using the loader tractor, we do what we can to toss about the materials as we assemble a pile. If bacteria-rich compost is required in very large volume, a compost turner may be worthwhile—though these are expensive specialty

FIGURE 10.5. Straw covering of compost piles gives them a skinlike protective layer.

machines rarely in use on smaller vegetable farms. We do make many of our specialty composts—the potting soil compost, vermicompost, and some of the side-dressing composts—by hand.

Once the initial pile is assembled, a coating of carbon-rich materials is applied over the surface of the pile. This is similar to mulching in the field in terms of how it protects the pile from drying winds and sun. It holds in moisture and allows for rain penetration as well as keeping down any potential smells. We often use straw or leaf for this covering, because wood chips on the surface of the pile would be less likely to begin decaying. We prefer an organic mulch coating instead of plastic covers. We bring out the plastic covers only during times of excessive precipitation.

Once the pile is assembled, it is time to apply Biodynamic compost preparations. These materials, though minute in volume, have the potential to greatly assist in the balance of the forces associated with the composting process and the succeeding crop production. Along with the 500 and 501 field sprays, the compost preparations were a gift by Rudolf Steiner to agriculture. An understanding of the reasoning behind the making of these compost preparations enhances insights into greater environmental conditions as well. Rudolf Steiner's agricultural lectures as well as Biodynamics in general may well be worth studying in this

FIGURE 10.6. Loaded wheelbarrows ready to roll down the wheel track. We never lift the compost, only dump and spread it out.

regard. The Biodynamic compost preparations are various herbs prepared inside various animal-based materials. The preparations are inserted into six holes placed on the side of the pile, one preparation in each hole. They can be produced on farm or purchased from reputable suppliers. Hugh Courtney at Earth Legacy Agriculture has supplied us with most of our preparations for many years.

Successive turning of the pile allows the producer to manipulate the compost further. The turning will introduce air and its drying influence as well as further homogenize the pile. It also gives the grower the opportunity to introduce more materials such as to adjust the C:N ratio or make further mineral adjustments. Turning generally accelerates the compost process by introduction of more air and the subsequent heating from increased decomposition. Turning encourages a bacterial decomposition whereas less turning is more amicable to the fungus and other organisms such as earthworms, which prefer cooler temperatures and less disturbance. The turnings are related to the air:water ratio balance of the compost pile, and again this balance is a primary area of concern for the compost producer. If more water is needed, then it will need to be added by irrigation. If more air is needed, then turning or covering with tarps to keep off rain may be appropriate.

FIGURE 10.7. Dump truck delivery of compost to potato hills. The truck fits over two 34-inch (86 cm) potato rows or one 58-inch (1.5 m) bed.

The need for compost maturity is variable, as discussed earlier. In general if the area where compost was applied is to be seeded or transplanted into immediately, then the compost pile will need to have cooled to pretty close to ambient temperature, maybe around 90°F (32°C) or so maximum. Hotter materials may be spread on the surface, allowed a few days of cooling and assimilation, and then seeded or transplanted into.

There are many means of compost spreading; this is a prime area to mechanize if the tonnage is large. The loader tractor is again critical to loading the material into a manure spreader, dump truck, or wheelbarrow. Some growers even spread directly using the loader tractor when appropriate for the bedding layout. Use of this equipment essentially eliminates the need for the grower to ever have to lift the compost, and instead requires only the manual work of spreading out the compost with rake or pitchfork to ensure even coverage. Compost spreading is very similar to mulch spreading, as described in chapter 6. We often dump compost into a line of wheelbarrows with the loader. This can be very quick, and the wheelbarrows often make for accurate spreading. One person can be driving the tractor back and forth to the pile, loading the wheelbarrows at the field edge (have some extra wheelbarrows available), while others are accurately spreading materials on the beds. Manure spreaders and dump trucks are used as well. In the case of spreading one bed at a time accurately with the manure spreader, the distribution paddles can be

POTTING SOIL INGREDIENTS

15 gallons (57 L) compost (prepared as described in "Tobacco Road Farm Compost" on page 177)

15 gallons (57 L) coir

10 gallons (38 L) perlite (optional)

5 gallons (20 L) rice hulls

5 gallons (20 L) worm castings or worm-aggregated topsoil

3 quarts (3 L) seawater (or 3 fluid ounces Sea-90—liquefied sea salt)

2 quarts (2 L) greensand

2 quarts (2 L) IMO (indigenous micro-organism culture; see chapter 11)

2 quarts (2 L) ground eggshells or diatomaceous earth

1½ pints (720 ml) dolomite limestone

1 pint (480 ml) dried seaweed

1 pint raw milk

1 pint homemade liquid fish extract (see chapter 9 for instructions)

8 fluid ounces (240 ml) Azomite rock powder

4 fluid ounces (120 ml) zinc sulfate

2 fluid ounces (60 ml) manganese sulfate

disconnected and limiter boards installed if necessary. Compost rates are of course variable upon materials and desired outcomes. Hopefully the previous chapters have provided an understanding of how to determine appropriate amounts. We spread compost before most crops, often about 30 tons of compost to the acre (74 metric tons per ha) or more, or about one wheelbarrow to 150 square feet (14 sq m).

Composts can also be prepared for side-dressing growing crops, as described in chapter 9. Side-dressing composts are generally prepared a month or more before usage to allow materials and nutrients to integrate, though they can be mixed closer to usage if needed.

Potting soil is another prime area of usage for composts. In this case the compost generally needs to be thoroughly decomposed and not "too salty" in terms of fertilizer inputs. Compost-based potting soils are well known for their ability to suppress diseases and obviously offer much more biological activity than a sterilized potting soil. When we make a compost for potting soils, the primary ingredients are cattle manure mixed with hay and leaves along with some other quick-decomposing materials for diversity. We do not add wood chips to this pile. If using topsoil in a potting soil recipe, it is labor saving to use soil from an area that does not contain large volumes of weed seed.

The compost pile for this mixture is relatively small, about 3 cubic yards (2.3 cu m) at assembly in spring, and is turned by hand several times before final assembly in September. At that time we repeatedly blend this compost with the other ingredients listed in the "Potting Soil Ingredients" sidebar to make a mixture that works well for soil blocks as well as in pots. The Biodynamic compost preparations are added upon final assembly of the pile as well. After the fully assembled pile has set for a month or so, we can screen it through a ½-inch (1.3 cm) mesh to sift out any chunks and debris as well as to further mix. The pile is then allowed to sit directly on the ground surface, exposed to rain and atmospheric conditions until mid-fall, when the pile might begin to freeze. We then cover it with a plastic tarp with a few holes in it to keep off excess winter snow and rain but still allow a little air and water exchange. A cover of straw about 1 foot (30 cm) thick is spread over the tarp to insulate the pile and keep it from freezing over the winter. This allows us to uncover part of the pile in February for use, as we start greenhouse seedling production at that time. We are still refining our recipe, but the list in the sidebar is what we use at present. The perlite is a new addition and is still experimental at the time of this writing. So far we don't find much difference in results based on the addition of the perlite. This potting soil produces stout, flexible, yet vigorous seedlings. Due to its high-carbon materials, it does require liquid nutrient additions, particularly calcium and nitrogen, for some seedlings. Milk, molasses, and liquid seaweed are common fertilizers for this. We repeat this recipe about 40 times to make our potting soil for a year's usage. This recipe is based upon our conditions, and growers may need to adjust the formula to suit their situation.

Vermicomposting is the use of large numbers of earthworms to accomplish the primary decomposition of materials. Worms do not tolerate temperatures much above 90°F (32°C), so this is a cooler process than regular composting. Earthworms like to escape into the general environment. To keep them concentrated in a vermicomposting system, we make bins out of plastic 55-gallon (210 L) drums. We cut away about a quarter of the drum and then elevate the open drum on a wooden frame placed under a shade tree. We use pieces of plywood as lids, as shown in figure 10.8. In the winter we cart these drums down into one of the root cellars to await warming spring temperatures.

We gather the worms to start and feed them on a mixture of partially decomposed vegetable scraps from the pre-composting stockpiles, along with the accompanying leaves and straw. We add some calcium and mineral grit materials like coarse-ground oystershell. The worms also seem to enjoy raw milk occasionally. Sometimes we add a specific mineral or seaweed as a way of incorporating a bioactive form of these minerals into our liquid extract side-dressing formula (which is made using a liquid extract of our vermicompost).

The worms are fed by piling fresh materials on top of the existing material. The worms move up into this newer material, which allows us to eventually

FIGURE 10.8. Vermicomposting bins made from 55-gallon plastic drums. During the warmer months they are kept in the shade of an old apple tree.

remove them by skimming off the top few inches of bedding. Then we can collect the material in the bottom of the drum, which we use primarily in our potting soil. We then return the worms to the barrel and the process starts over.

Each drum has a 2-inch (5 cm) diameter hole drilled into the bottom bin with wire mesh window screening caulked in place over the hole. This allows excess moisture to drain through the bedding. When we require a compost extract for buffering our side-dress fertilizer, we water the bins (about an hour before the liquid is needed). We place a bucket below each drain hole to collect the liquid vermicompost extract. Using the liquid soon after extraction is important. If allowed to sit the extract may develop detrimental anaerobic compounds.

When discussing composting with other farmers, one recent concern that frequently arises is phosphorus levels. Various laws and regulations limit the application of phosphorus-containing materials, including compost, in the name of water quality. This kind of broad sweeping regulation defies common sense and agricultural tradition and experience, stands on weak science, and is unlikely to help protect water quality when the greater picture is considered. That said, phosphorus can be high in composts and if applied to soils that also have high phosphorus levels, it can lead to excess. Phosphorus levels are closely related to the grain ration in animal feeds, so usually the manures of grain-fed animals are responsible for elevated phosphorus levels in compost. We don't use manure from grain-fed animals in our compost, and we have seen the phosphorus levels on the Mehlich-3 soil test drop even though we apply 30-plus tons of compost per acre per year. See chapter 9 for more on phosphorus levels.

CHAPTER 11

IMO: Indigenous Microorganisms

As part of our efforts to improve the biological condition of the soils and the quality and yield of produce, we make and use a biological inoculant that contains large volumes of active, locally sourced microorganisms. This inoculant material is referred to as IMO (indigenous microorganisms). The techniques for IMO culturing come from the principles of Korean Natural Farming (KNF). We were first introduced to Korean Natural Farming around 2010 when a farm intern returned from Hawaii where she had been observing its usage on farms. There is a significant population of Koreans in Hawaii, and the techniques of KNF have transferred over to the English-speaking population there. Many practitioners in Hawaii, along with the extension service, have helped to disseminate KNF agricultural techniques to mainland America.

The indigenous biological condition of soils is altered and usually damaged by agricultural methods and materials used in vegetable culture. This includes tillage, various fertilizers, and pesticides. In addition, pollution and climate manipulation combine with these factors to limit soil biological function and development. To reintroduce local microorganisms from less disturbed environments that have been brought to a high level of activity through the culturing steps delineated here can dramatically improve growing conditions. This improvement is similar to the renowned growing conditions of a "new" field, one that has been growing a mix of perennial vegetation for an extended period, when plowed and prepared for vegetable growing.

The IMO process involves setting a box of partially cooked grain into an appropriate place in the environment in order to collect a culture of that local biology. This culture is then brought to the farm and put through various processes that result in a substantial pile of highly active local biology. This material is then introduced into the farm system via irrigation, foliar sprays, potting soils, live-stock water, mulch piles, and direct placement onto the fields. The culture is full of diverse, well-adapted organisms that will continue to thrive under field conditions if treated carefully. No-till is a primary approach to continue the proliferation of the IMO under field conditions, but avoidance of life-damaging chemicals is important, too. The addition of methods and materials that enhance soil life perpetuates the IMO. Due to pollution and the unavoidable soil damage of some aspects of vegetable growing, however, intermittent reintroduction of IMO to the field can be beneficial. The mycelium of fungal organisms is obvious during the culturing, and when IMO is introduced into a field, the mycelium of the fungal organisms can be observed proliferating. Mushroom development may follow. This introduction is a sign of successful integration. IMO have aided in the extensive development of soil aggregation and improved soil structure in our fields. Soil and tissue test results show improved release of nutritive elements such as calcium, magnesium, and phosphorus as well as micro and trace nutrients, likely due to enhanced fungal activity.

Making IMO #1 and IMO #2

Producing IMO cultures is a four-step process: inoculating a grain media with indigenous microbes; incubation of the culture mixed with sugar; proliferation of the culture in a bran substrate; and addition of soil to the culture. The grain

FIGURE 11.1. IMO #1 just retrieved. Leaf litter was placed directly on top of this grain to assist the culture.

medium for IMO is usually brown rice, though we have used other grains as well. It is important that the grains do not contain biologically damaging pesticide residues to any significant extent. The rice/grain is cooked with water at a 1:1 ratio by volume for about 20 to 30 minutes. The result will be a firmer, drier rice than would be served for a meal. The cooked rice is then placed in a container such as a wooden box or basket. We use a box made of cedar (9 × 16 × 6 inch / 23 × 40 × 15 cm interior dimensions) with a lid. Cedar is the preferable material, but we have used pine boxes. The box is filled about two-thirds full, leaving space for air. A lid or other covering is essential in order to keep out excessive rain while the box sits in the culturing environment.

The box is then taken into nearby forest where an appropriate culture site is located. Here is a time we fully use our senses in order to locate a site. The signs for a site often come from the smells of the forest area, as well as using our sight and other senses. Often the location is under a large deciduous tree that has shown its strength over time and fed a strong biology at its roots. When a site is located, the forest duff is then moved aside and the box is set down. The duff is piled atop the box until it is fully buried. Sometimes we place exceptionally vigorous-looking and smelling duff directly on top of the rice in the box, but this is not necessary.

The box is left buried for approximately one week—longer in cold weather, shorter in hot weather. After this time, the box is unearthed and the contents are examined. If culturing proceeded well, a rich white fungal growth on the

FIGURE 11.2. IMO #2, which is sugar and IMO #1 mixed in a crock. We cover the crock for storage in the root cellar.

grain—with perhaps a few growths of blues, reds, greens, and the like—is present. If there is less white fungal growth than growths of other colors, the options are either to take only the part that is mostly white, or to start again. We have never had a box fail completely. This culture is called IMO #1.

We take the IMO #1 back to the farm and mix it with a pesticide-free brown sugar at a 1:1 ratio by weight in a clay crock, leaving about one-third of the volume of the crock for air. The crock is then covered with a porous paper material secured by a large rubber band. The crock sits for approximately one week at a warm room temperature, after which it is ready for use. This material is IMO #2. It may also be stored for later use by placing the crock in a root cellar; alternatively some people store this material in their refrigerator.

Making IMO #3

The next stage is the making of IMO #3. For this step, IMO #2 is liquefied by stirring in water, then mixed into a pile of bran. The bran is piled on the ground in partial shade underneath deciduous trees. The bran pile is hydrated to about 65 percent moisture, or until water can barely be squeezed out of a ball of the bran, using a dilution of IMO #2 at 1:500 water or so. For this step we mix a fluid ounce or two of IMO #1 into 4 gallons (15 L) of water, stir, and pour over the top of the bran pile. Other KNF herbal extracts are often added at this point, such as garlic, ginger, angelica, licorice, and cinnamon extracts, at 1:1,000 dilution; also added are sugar extracted fermented plant juices (FPJ) as well as vinegar at 1:500 dilutions. These extract materials are very useful components of the KNF farming method; their making is described in chapter 9. IMO #3 can be made without their use, however.

The height of the bran pile is often only 3 to 4 inches (7.5–10 cm or less), and hydration is accomplished with turning for consistency of moisture. The height of the pile depends on how high the ambient temperature will be in the next few days. Lower piles tend to heat less; higher tend to overheat. The optimal temperature range is about 100 to 120°F (38–49°C). If the pile starts to heat much over 120°F, we turn it in order to cool it. We cover the pile with wet leaves, straw, or cardboard and protect it from excessive rain. We often use a black plastic tarp suspended on hoops to keep off rain and provide more humidity.

The most common bran used in the United States is wheat bran; in Korea rice bran is utilized. (Rice bran is probably quite plentiful and inexpensive in Korea.) Bran that does not contain detrimental pesticide residues is best. These brans are highly valued in the United States for use in animal feeds so they are relatively expensive. We have found a sifted floor bran from a local grower of wheat at a very reasonable expense, but often we have to look to commercial sources. We make several piles of IMO a year, using 200 pounds (90 kg) of bran for the IMO #3 stage, at a cost of up to $100 to $150 per pile. This generates enough for our 3 acres (1.2 ha) of vegetable crops. We have used rice, wheat, and oat bran and have had success with all of them. Oat bran is more expensive than other brands, but its gummy moisture-retaining nature seems to best assist fungal growth. We generally pile the oat bran very thinly on the soil, 1 to 2 inches (2.5–5 cm) thick. We have exper-imented with many less expensive substi-tutes for the bran, but none of them have given us the consistent results of bran.

FIGURE 11.3. A frothing IMO #3. Frothing is a useful indicator of life.

The pile of bran quickly begins to heat and develops quite fragrant odors. Within days a white fungal growth very similar to the IMO #1 culture covers the pile. In drier times of year, or if the pile overheats and requires turning, we add some water to rehydrate the pile. This is often after three or four days. The explosive growth of the biology does quickly consume moisture. Once the fungal mycelium has thoroughly grown into the bran, often after only five days or so in the summer but longer in the cooler periods, it is time for the incorpo-ration of soil and the assembly of IMO #4.

Making IMO #4

To develop the IMO #4 stage, soil is mixed into the highly biologically active bran pile (IMO #3) at a ratio of 1:1 by volume. The soil can come from a variety of places, and diversity is probably of benefit, though it seems likely that soil from the area to be farmed is best. We use a mixture of topsoil from the field, topsoil from the field edges, and field subsoil from various digging projects. The soil helps to cool the bran pile if it is still heating excessively, so the material can now be piled higher, usually about a foot (30 cm) tall.

The highly activated biology of the IMO #3 now has a chance to incorpo-rate into the soil and continues to remain highly active, though at generally lower temperatures than the straight bran. The clumps of bran formed during the IMO #3 process are best left somewhat whole during this mixing process, usually about the size of golf balls (1 to 2 inches in diameter). The pile is rehy-drated to a similar level of hydration of IMO #3 using herbal and plant extract dilutions if available, as well as seawater at a 1:30 dilution (also if available). After about a week the material will be ready for use. It is most active for a few weeks after final assembly, but we store this material by simply leaving it under its tarp for up to several months and apply it as needed.

We apply IMO #4 directly to the bed surface before seeding or transplanting, at a rate of about 5 gallons to 250 square feet (0.8 L per sq m), more or less depending upon availability. IMO #4 is capable of expanding itself in field conditions and is like a catalyst, so applying exact volumes may not be critical. Under gentle treatment the biology may persist for extended periods so intermittent field treatment with IMO #4 may be sufficient, depending on individual conditions. We do not treat every bed every year this way, but we do treat most beds at least every other year. It is best to apply IMO in the morning or evening, cover it with mulch, and water it in. The IMO can be readily seen expanding into these mulch materials (straw, leaves, wood chips) following this process. At the rate that we apply this material, we can immediately seed or transplant, but with higher applications a few days of cooling down may be appropriate.

As described in chapter 9, another useful application of IMO #4 is foliar application. For this process approximately 2 gallons (7.6 L) of IMO #4 are vigorously stirred into about 5 gallons (18.9 L) of water, then strained and added to about 25 gallons (94.6 L) of water containing other microbe-friendly foliar ingredients. This is then immediately applied. IMO can also be applied using an irrigation system with a similar approach of stirring and straining. We have also used IMO to reduce odors in animal bedding areas. (Livestock particularly enjoy consuming IMO, so the culturing of IMO #3–4 may need to take place in a fenced-in area to keep animals at bay.) Occasionally we add it to animal drinking water to help digestive functions. In orchards IMO can be added to the mulch that is applied under the trees. As well, we often utilize the trees' shade for the actual culture area of IMO #3 and #4, which is directly of benefit to the orchard. It is also an ingredient in our potting soil formulation.

Much experimenting can be done with the IMO process, and the material has many uses. Often we mix multiple cultures of IMO #2 into the water to hydrate the IMO #3, giving greater diversity. Taking IMO #1 from various sites and at various times of the year aids in the diversity of organisms as well. The use of various materials in the culture steps of IMO can

FIGURE 11.4. IMO #4: Soil has been added, the temperature has cooled, and fungal growth continues.

FIGURE 11.5. The first sugar maple (*top*) resembles the others in our region with its sparse canopy, small leaves, and yellow-green coloration. The second sugar maple (*bottom*) has had IMO#3 and #4 cultured underneath several times. When we started IMO production, this tree looked worse than the no-IMO tree. Now with its full canopy of dark green leaves, it resembles the trees of old.

also yield a dominance of certain types of organisms; these could be mixed with various aims in mind, such as more carbon-rich materials for fungus or nitrogen-rich material for bacteria. IMO #4 can also be used in the preparation of various side-dressing fertilizers, liquid or solid. People have experimented with making cultures from virgin forest soil and bringing back that biology to their farms. Another common approach to save on expense is simply to utilize IMO #2 directly in foliar sprays and irrigation, thus eliminating the bran culture. We have not tried this, preferring to fully develop the culture on bran and soil. IMO can also be cultured in greenhouses during the winter. The methods put forth in this chapter are certainly not the only way to culture local microbes; hopefully they provide helpful guidance.

CHAPTER 12

Weed, Insect, and Disease Control

M anaging weeds, insects, and disease can be a serious drain on profitability, because it is work not directly related to the growth, harvest, and sale of a vegetable. Much of this manual has sought to describe methods that will help growers encounter fewer difficulties with pests by development of systems that increase the health and vitality of the crop. Those systems are of fundamental importance. They have maintained our farm in a constant state of profitability without any need for pesticide application. However, there are always a few pest issues that still need the special attention of the grower even in a highly functioning system. This chapter provides ideas and concepts for growers who still may face many of these difficulties as they are moving their farms and gardens to more balanced conditions, as well as for growers who face only occasional difficulties.

As growers we can do much to help create balance in our soils and crops, but often greater environment disturbances can come in and upset this balance. This is happening with more and more frequency as detrimental human influences spread over our farms and gardens, often causing crop failures, frequently from pestilence. Growers need to nurture strong, well-balanced systems to resist these disturbances. This interaction between the influences resulting from the whole of human activity and the crops greatly impacts human civilizations at a macro scale as well. Desertification, famine, and exodus are prime examples. The field is a mirror of the grower's interaction with the various forces at work in the environment. In this sense weeds, insects, and disease are beneficial: useful indicators offered by nature to help growers adjust their practices. They can also be beneficial in that they may destroy crops that were severely out of balance; crops that, if consumed, could bring their imbalances into the consumers' bodies as well.

Weeds

Weeds can be particularly beneficial because they cover the earth with a living canopy of leaves that collect solar energy and therefore create sugars and other compounds through photosynthesis and its results. These materials then are fed into the soil ecosystem, further enlivening it. "Weed" is a human concept: a plant that is out of place. Actually, weeds don't exist in nature, because all plants benefit the soil life and help develop the soil toward greater ability to sustain life. Nature utilizes whatever plants are available to improve conditions for life. At this point in the earth's evolution, with the damages wrought by so many detrimental human activities, nature often utilizes very fierce, aggressive, fast-growing plants in this effort to protect life. It is not a random occurrence that heavily disturbed areas are full of poison ivy, prickers, and other tough plants. They help to keep detrimental humans away from the land so disturbed, to allow it time to heal. Similarly, in agricultural fields, the more detrimental the growing practices are, the more aggressive the weed problem. The gentler and more nature-friendly the agricultural practices are, the less weeds pose a problem. We have repeatedly seen this in the trials of agricultural practices and materials at Tobacco Road Farm, and it is an important component of why we have shifted to a no-till system.

Weeds in a field are particularly useful indicators of what the field is trying to tell the grower. Growers will benefit from a study of weeds and their manner of growth to learn which grow best in specific soil conditions. Certain weeds grow well in soils that are compacted, while others do better in soil conditions that are wet, acid, alkaline, anaerobic, excessively aerated, high in calcium, low in calcium, high in nitrogen, low in nitrogen, and so on. Weeds are capable of showing growers what the characteristics of the soil air and water are, what nutrient deficiencies/excesses exist, the effects of tillage and crop rotation, and much more.

Weeds are highly adaptive, and species dominance can change quickly in a field as a grower changes practices. A common example of this in vegetable growing is the shift of weed species to fast-growing weeds such as galinsoga. Galinsoga can produce viable seed in as short a time as a few weeks. It thrives on soil disturbance and often becomes the dominant weed in vegetable systems that can effectively control weeds that are slower to mature. Galinsoga was our primary weed when we used tillage and was extensive in its distribution. With the switch to the no-till system, galinsoga germination dropped to less than 1 percent of its previous numbers, because we no longer provide the conditions for its proliferation.

Weeds also work together as a group in that they will signal to one another when to germinate. In this case all the seeds that could germinate in certain conditions may well not. Instead the appropriate number of weeds will germinate to cover the soil, then these germinating plants signal to the other seeds to wait until they are needed in the future. This is sometimes referred to as allelopathy, but is more than a simple chemical suppression of growth—the plants are actively communicating with one another. Weeds also may either inhibit or enhance the growth of neighboring plants through their various root exudates and other characteristics. Often young weeds can enhance the growth of a neighboring crop by assisting in covering the soil with growing plants and thus creating a better-functioning living system.

REDUCING THE WEED SEED BANK

The overall volume of weed seeds present in the soil of a field or other area is often called the weed seed bank. As growers our practices can either increase or decrease this bank, and unfortunately it is much easier to dramatically increase it than to decrease it! Appropriate practices can steadily reduce this bank. No-till offers much in this regard, with appropriate fertilization, and weed control tools also contributing their part. The objective is to not allow weed seed to fall upon the soil in the growing area. One factor in the growers' favor is that weed seeds do eventually deteriorate in the soil if conditions do not present themselves for germination.

To reduce the weed seed bank, growers can destroy weeds as they germinate and create conditions where weed seeds decay rather than germinate. Seeds of some weed species decay in the soil faster than others, but time is on the grower's side as long as the seed is not exposed to conditions for germination. This is an important benefit of this no-till system: The soil environment is unfavorable to weed seed germination. Increased soil life helps to speed the decline of the seed bank, too, as earthworms and various other organisms consume weed seed in the soil. Seeds left on the surface of the soil are vulnerable to being eaten by insects, birds, and rodents, which can greatly reduce their number. This often occurs in the fall and winter after annual weed seed has fallen to the ground and before spring germination. Of course, this would mean that it is generally best not to mix the weed seed into the soil, as often happens when growers perform tillage events after harvest.

One difficult situation that is not easy to remedy is the introduction of new weed seeds through materials brought in from outside the farm. As pointed out in earlier chapters, compost ingredients and mulch materials from outside sources need to be carefully evaluated to avoid bringing in problematic weeds. Composting may be able to destroy some weed seeds during its decomposition cycle but this is often incomplete, so though a few weed seeds may be tolerable, off-farm materials with much seed may be best left alone. This frequently happens off-farm where these materials are piled and either weed seed is picked up when collecting materials or weeds are actually allowed to grow upon the piles. This is especially common on areas where manures are piled. Other sources of weed seeds can be contaminants in crop seed, cover crop seed, or animal feeds.

Rotating crops from year to year can aid in weed control by varying the growing environment. This limits the potential of specific weeds that proliferate when certain crops are grown. Crops with limited plant canopies, such as the alliums, are prone to weed difficulties because they have a limited ability to shade out competing plants. Crops like squash or cabbage with vigorous leaf growth can actively shade out all but the fastest-growing weeds. Buckwheat and some other cover crops are also capable of this smothering and shading effect. Thus alternating these fast-growing crops and cover crops in a crop rotation can have a positive impact on overall weed control. Rapid crop turnover can also reduce weed competition. A crop of red radish or arugula can be harvested in about 30 days, for example, and few weeds are capable of producing seed in this short a time period. (Galinsoga is a noteworthy exception.) After harvest, preparation for the next crop, including mowing and solarizing, should eliminate any weeds still growing. Therefore including short-season crops in a rotation plan can be very effective at reducing the weed seed bank.

When using cover crops for weed smothering, appropriate seed rates for high density are needed. It is also important that weeds are not able to reseed themselves during the growth cycle of the cover crop. If a cover cropped area develops weeds that are approaching seed formation, the cover crop would need to be terminated with appropriate reseeding or covering to follow. If weed control is already well in place, cover crops can be interseeded into growing crops. This can be helpful in improving plant canopy coverage of the soil along with furthering weed control.

Nature will always seek to cover bare soil with vegetation such as weeds, so growers can assist nature in this effort by either heavily mulching the bare soil between crops or growing a cover crop in these periods, and maybe even a bit of both. During a cold time when cover crops will not be able to effectively grow and cover the soil, the grower's best option may be to lay down a thick coating of organic mulch materials. If there is enough time and temperatures are appropriate, however, growing a cover crop is preferable for not only weed suppression but also soil development.

Tillage has been discussed as having significant impacts on soil air:water and soil structure, but the ramifications of tillage on weed germination are also very important to consider. When a soil is tilled, nature senses an extreme disturbance, which must be responded to with as much regrowth as possible in the form of weeds. The incorporation of air, exposure to light, and churning of the weed seed bank are just some of the explanations for this massive weed response. This being said, tillage can be carefully utilized after a flush of weeds has germinated, to destroy this flush and reduce the weed seed bank. This can also be accomplished in a gentler fashion with solarization events. Either way, the concept is the same: If the soil surface has been prepared through tillage, allow weeds to germinate, then destroy them, possibly even in successive events with breaks to allow additional weed flushes to germinate. This can also be achieved with the use of sharp wide hoes, rakes, or other weed control tools. This technique is sometimes referred to as fallowing for weed control. Using tillage equipment like rototillers and field cultivators in this manner can be very effective for the destruction of tenacious perennial weeds in particular, though it does come with the associated destructive impacts on soil quality.

The surface application of compost and mulches greatly reduces the capacity of weeds to germinate, particularly if they are heavily applied. This is a very useful approach when first moving a field into no-till, especially if that field is infested with a large bank of annual weed seed. The smothering characteristic of the materials is one aspect of this, and the positive effects of the materials to soil fertility aids the crop in establishing and growing quickly. This ability of crops to spring right out of the ground at a running pace is particularly useful for later weed control, because the crop will be larger than the weeds. For this strategy to be most effective, the crop seed needs to be of highest quality, planting or sowing should be performed soon after final bed preparation, and soil fertility should be well balanced. Immediate irrigation after seeding, as well as adding crop stimulant materials such as seaweed, seawater, vinegar, and plant extracts to the irrigation water, also can aid quick germination and early growth. Setting transplants rather than direct seeding can also greatly assist later weed control because of the difference in the size of the crop compared with the weeds, which will not even have germinated at the time the crop is set.

DESTROYING WEED GROWTH

Once a bed is shaped and the crop seeded, various tools and techniques can be effective to destroy weed growth, including hoes, flame weeding, solarization, and mulching. Hoes of many shapes and sizes are pulled by hand, pushed (like a wheel hoe), or mounted on a tractor. Hoeing is a skill to be acquired. Speed and accuracy are constantly challenged, and thus hoeing requires not just physical strength but sharpness of mind, and especially speed of decision making to proceed quickly. If these conditions are met, hoeing can be very fast and effective.

The usual objective of hoeing is to sever the top from the root of the weed, though some burying of small weeds also may be achieved. Most hoe blades travel about ½ inch (1.3 cm) below the soil surface when properly employed. It is more effective to hoe on a sunny day, because cut weeds generally are left to die in place, and the sun's heat kills the cut weeds quickly before they reestablish

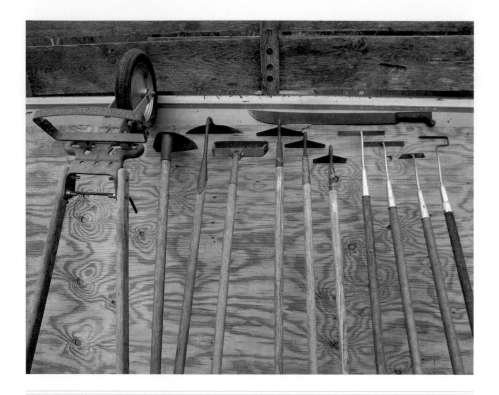

FIGURE 12.1. A collection of weed control tools. *Right to left*: wire hoe, three collinear hoes of different widths, three standard hoes of different widths, a scuffle hoe, a wide swan neck hoe, a grub hoe, and a wheel hoe mounted with a 12-inch (30 cm) scuffle hoe blade. A machete tops them off.

roots. Hoes are best kept as sharp as possible, usually by use of a grinding wheel. The blades also are best if kept rust-free for a smooth passage through the soil. Another hoeing technique is to throw soil onto the row to smother weeds that are smaller than the growing crop.

Straight, evenly spaced rows make hoeing easier and more efficient. If the rows are misaligned and the space between rows becomes too narrow, a hand-held hoe can simply be rotated to have a narrow profile. This is not an option when using wheel hoes and tractor-mounted hoes, however. Mulches and crop residues do get in the way when hoeing, slowing the process. However, with some skill, hoeing can effective to some degree in mulched soil conditions if required.

Mechanically it is generally easier to pull tools through the soil than push them. Wheel hoes must be pushed, but they can be very effective in areas where the soil is loose with minimal residue and the weeds are relatively small. In these conditions pushing a wheel hoe can proceed very quickly with minimal effort. Weed control implements mounted on a tractor can produce results very fast, but only if the field

layout has been well planned to accommodate the machinery. Adjusting implements precisely can take much time; to be efficient, large areas need to be laid out
with very even row spacing. It also is very easy for tractor-mounted tools to over-
work soils unnecessarily, a condition that rarely happens with hoeing. Our use of
tractor-mounted implements for weed control is negligible nowadays because of
the low weed pressure on our farm and the wide diversity of crops at spacings to
take advantage of highest yield and canopy coverage. Once seed has been placed in
the soil, the first opportunity for weed control is pre-emergent weeding. This form
of weed control is particularly useful in slow-germinating crops like the umbels
(carrot, parsley, and parsnip) and is not practical for fast-germinating crops. In the
case of carrot seed, which often takes 10 days or so to germinate, weed seedlings
may be dispatched on day 7 or 8. In this way the first flush of weeds is destroyed
and the carrot seedlings emerge within a couple of days into a weed-free bed. If
further hoeing is planned and the crop was seeded in furrows, it is important
not to disturb the seeds from their furrow spacing. Because of this, shallow pre-
emergent hoeing is carefully done parallel to the rows and thoroughly over the
entire bed surface. Wide hoes are utilized to achieve this. Solarization (as described
in chapter 4) or flame weeders can also be carefully used for pre-emergent weed
control. Flame weeding is detrimental to organisms that live on or near the soil
surface, so we generally no longer utilize this approach. We have solarized carrot
beds before germination for pre-emergent weed control with great success, but
this does require appropriate weather conditions.

After seedlings of a broadcast seeded crop emerge, surface weed control
options are basically reduced to either finger-and-thumb pulling, or cutting
off individual weeds with a knife. These approaches also work well for in-row
weeds in line-seeded beds. (Hoeing is faster for weeds between rows.) Finger-
and-thumb pulling of weeds is not appropriate if pulling out the weeds would
disturb the crop roots. Once annual weeds reach that size, they are cut off at
the soil line with a serrated knife instead. One advantage of finger and thumb
weeding or knife weeding is that the weeds are in hand so they can be collected
and removed from the field relatively easily, assuring that there will be no
regrowth. Perennial weeds require special consideration—cutting them with
a knife is not sufficient for termination, though it can delay seed formation.
Perennials have to be pulled out either if their growth will significantly reduce
crop yield or if they will further proliferate by going to seed or detrimentally
expanding their root mass before the crop is harvested. If this is not the case,
then perennials are often more easily dealt with after crop harvest.

If a crop was seeded or planted into straight furrows, then hoeing can begin
after germination. Hoeing after crop germination is first done with very small
hoes like wire hoes that will not bury crop seedlings with soil. It then progresses
to thin-bladed hoes like the collinear hoe that pose little risk of smothering the

FIGURE 12.2. Annual weeds can be cut off using a serrated knife just below the soil line with no soil disturbance or disruption of the growing crop.

crop. Later hoeings are achieved with bigger hoes that throw soil to the side as they are pulled and may progress to hilling hoes and equipment meant to throw relatively large volumes of soil toward the crop line. Because vegetable crops are grown at a wide range of between-row spacings, it's useful to have hoes of many sizes so the entire width of the between row space can be hoed in a single pass. This can be achieved by buying wide-bladed hoes and grinding or cutting the blades to the length of the between-row spacings.

Mulching or further mulching row-planted crops after the use of weed control tools can provide additional weed control and soil benefits. If weeds are growing amidst a crop that is close to maturity and the weeds are threatening to go to seed, a machete works well to chop off their tops just above the tops of the crop plants. If the weeds have not produced mature seed yet, then the cut seed heads can be left in the field. But if mature weed seed may be present, the seed heads are gathered and removed from the field. If the weeds are low growing, they can be hoed out and gathered up with pitchforks. The weeds can also be mown and bagged with lawn-mower-style equipment. The critical goal is to ensure that no viable weed seeds are allowed to fall to the ground in the field.

Pests and Diseases

Much of this manual focuses on techniques growers can employ to raise crops that are inherently resistant to insect pests and disease. In this regard, the importance of the balance of fertility and environment cannot be overstated. The grower can learn to recognize the physical signs of a well-grown, balanced crop by closely examining the appearance of crops that lack insect and disease problems. Observe leaf colorations and thickness, root growth, characteristics of balance, top and bottom growth. In sum, look for all the physical manifestations of healthy growth as described in chapter 2. With repeated study over time, the appearance of crop vitality and pest resistance becomes readily apparent. Include nighttime examinations, because some insects (both pests and beneficials) are primarily active at night.

LETTING NATURE BE YOUR ALLY

If problems arise, it is important to remember that we are not alone in our efforts to rebalance growing systems. Nature has many mechanisms for reestablishing balance, providing growers with many allies. If an imbalance brings multitudes of caterpillars, nature in time establishes animals and insects in proper proportion to consume this new potential food source of caterpillars. Similarly, if imbalance brings hordes of rodents, nature will supply the rodent eaters. If disease wipes out 99 percent of a crop, the 1 percent left will yield future plants that are more resilient.

As growers we can do much to assist nature in this effort to reestablish balance, leaving us in an ever-stronger position. It is important not to overreact when the wave of imbalance is upon you. Growers commonly attempt to counteract the damage, getting in the way of nature's efforts. A simple example of this: employing an insecticide when insect pest populations have already reached high levels and are causing severe damage to a crop. When pest populations reach a high level, this is often the point at which beneficial insects are establishing as well. The insecticide can destroy both prey and predator, thus leaving the cycle to continue without natural control ever building up. Over time, we have seen many pest imbalances wash over our region. Though the infestations were certainly damaging at the time, it was uplifting to witness the response of nature, and to be of assistance in the natural rebalancing process provides hope for the future.

One early indicator at Tobacco Road Farm that gave us a dramatic insight into how plants naturally resist insect assault was flea beetles on brassica crops. When brassica crops were fall-sown and overwintered, they would be flea-beetle-free through their entire life cycle. We decided to run a trial, seeding the same variety of brassicas in the spring next to the overwintered ones. The spring-sown seedlings became heavily infested, yet the neighboring fall-sown planting remained completely flea-beetle-free. The fall-sown crop was at a different stage of maturity than the spring-sown, but that state of maturity alone did not explain the total absence of flea beetles, because the spring-sown

FIGURE 12.3. This arugula, which was seeded in early May and not protected by row cover, shows no flea beetle damage even after four weeks' growth in open air.

FIGURE 12.4. This is from a stand of 20,000 garlic plants. At harvest we found barely any disease or second-quality bulbs.

crop was still assaulted by flea beetles later in the season when it had reached the later stage of maturity. What this stunning example showed is that the influence of growing conditions can lead to complete resistance in an otherwise susceptible crop.

Following that simple early example, we have observed many other examples of crop resistance, particularly when crops were grown in balanced conditions. As our soils have developed and improved, especially since the shift to no-till techniques, many insects and diseases that were previously damaging to crops are no longer problematic. We can direct seed cucumbers, melons, and squashes in the field without any significant damage from cucumber beetles and other beetles or bugs that commonly plague these crops. Radishes, turnips, and rutabagas frequently show no symptoms of root maggots or black rot. Stem rot of garlic is almost zero out of the 20,000 plants we harvest annually. Flea beetles now are often completely absent, or we find only slight damage. Basil downy mildew does not damage crops until well into September. These are just a few examples of the many.

It may take some time for growing conditions to improve sufficiently on a farm to bring about such remarkable levels of insect and disease resistance.

However, it is important for growers to carefully look for even small reductions in insect and disease pressure, as these yield indications of what grower actions might be inducing improvement. Trials and actions can then be developed and further observed.

Soil fertility conditions not only impact degrees of resistance to insect and disease but also can increase growth rate and size of plants; this, when combined with high planting density, can diminish pest damage. This is the pest to crop ratio. For instance, if there are 50 mice in 1 acre, and a single 100-foot row of potatoes is planted, the damage may be significant. However if that entire acre is planted with potatoes, the damage caused by those 50 mice will seem minor. This is true of insects and many other pests as well. Large volumes of rapidly growing, strong crops simply dilute the level of damage. Growers often take advantage of this by setting transplants into a field rather than direct seedings. The transplants, comparatively much larger than newly germinated seedlings, will be better able to survive any potential pests.

ROTATING CROPS

Crop rotation serves many purposes, as described in chapter 7, including avoidance of insect pests and disease. Rotation by plant family can move crops away from areas where pestilence has built up. This is most effective if the crop areas are physically separated by a relatively long distance, at least as far as the pest insect in question will fly. If growing areas are sufficiently separated, it can be particularly useful to rotate the multiple plantings of a crop over the course of the year, or even year to year. For example, plant the spring cabbage in field #1 and the fall cabbage in field #2. Plant winter squash in field #1 this year, field #2 next year, and so on. Crop timing is also a way to reduce pest difficulties, especially if a crop is grown in its naturally favored season. This relates back to the flea beetle example described above. Flea beetles are generally not a problem when brassica family crops are only seeded in their preferred summer season, with no brassica seeded in the spring. Potato beetles will not be a problem in our area if we plant potatoes only in June, without earlier plantings in that field. Seeding or transplanting a crop at its most appropriate season can be very beneficial to its vigor and therefore its strength and resistance to pests. Of course if soil fertility conditions are particularly strong, any crop can be seeded at any time with little danger of pestilence.

See chapter 9 for a discussion of foliar application of materials that assist disease and insect resistance. Materials such as silica and vinegar along with the trace elements and beneficial biology supplied in this manner can be very effective. Many of these materials assist by balancing crop growth and increasing health; others by directly inhibiting the ability of the pest to proliferate.

BREEDING FOR RESISTANCE

Chapter 2 noted the importance of sourcing high-quality seed for insect and disease resistance. In addition growers can take advantage of breeding efforts to produce varieties that are naturally resistant to various insects and diseases. Breeders are actively trying to enhance such traits in many vegetable crops. However, growers themselves are often in the best position to develop crops with resistance to the pests and diseases in their specific environment. Breeding crops for improvement and to keep up with the changing environment is very important at this time. Growers can certainly select for the most vigorous, healthful, resistant plants in their environment, and save seeds from those plants in pursuit of constant, year-after-year improvement. They can also crossbreed plants to create similar conditions and thus benefit from genetic diversity and hybrid vigor. This can be particularly effective in the development of landraces.

ENCOURAGING BENEFICIALS

Another great advantage of seed saving is its positive impact on beneficial insect populations. Beneficial insects are assisted by the growing of seed crops in two ways: The crop provides an appropriate habitat for the insects' survival and can also provide a food source. When growers clear out crops immediately after reaping the marketable harvest, the beneficials often lack enough time to complete their life cycles before the field is renovated for the next crop. The lengthy growing season required to harvest a crop of mature seed crops provides a long-lasting field cover, which protects the beneficials as they complete their life cycle. The crop can supply predatory beneficial insects with appropriate prey as well, or pollen and nectar for pollinating insects. Some beneficials partake of both prey and nectar from the crop.

An example of a beneficial insect that is highly dependent upon this type of relationship is the parasitoid wasp *Cotesia glomerata*. These small wasps prey upon cabbage caterpillars by laying eggs in a host caterpillar. The young larvae then consume the caterpillar and emerge to spin small cocoons on or around the now dead host. The wasp then pupates in the cocoons and emerges as an adult wasp to begin this 20-to-50-day cycle again. The adult wasp feeds upon the flower nectar of cabbage family crops. In this way the cabbage family crop provides food sources for both the larval development and the adult wasp as well as the habitat and continuity for the completion of the wasp life cycle. By allowing the crop to flower and produce seed the beneficial is provided for, but what would happen if no cabbage family crops were to flower, or were always completely harvested and residues turned in? The *Cotesia* wasp overwinters in its pupal cocoon upon and often underneath the leaves of cabbage family crops. By allowing cabbage family crops to overwinter and flower, and thus produce seed, the life cycle of the *Cotesia* wasp is fully supported. We have witnessed this delicate balance of nature as the flowering of an overwintered cabbage family crop directly coincides with the spring emergence of the adult wasps—to the exact day. This is but one example of how seed production is integral to the welfare of beneficial insects and therefore of benefit to future crop generations. It points to the interrelationship between grower and crop and indicates that a co-evolution is appropriate. The crop is not to be exploited for human interest alone; rather it feeds us so that we can assist its own perpetuation through seed saving and associated selection and improvement. Even small areas of seed production can offer multitudes of beneficial insects, along with the potentially better-adapted seed.

Growers can support many beneficial insects and other animals by providing food sources and habitat. Various insects, birds, bats, snakes, frogs, toads, cats, and dogs are but a few examples. Tending flower gardens on the farm can be very beneficial in this regard. The diversity and general fertility of the surrounding landscape can also be very helpful in maintaining beneficial populations. The cultivated growing area is not an isolated entity; it is highly influenced by how the grower cares for the surrounding area. More diversity, more habitat, more flowers, better-balanced fertility in the environs can dramatically impact pest pressure. Firmly establishing that insect and disease difficulties are highly related to the growing environment, and the condition of a susceptible plant, growers' attitude toward the need for field "sanitation" often change. If there is no inoculation capacity for a disease in your region, or absolutely no pest insects of a particular species in your region, then your crops will not be infected by that disease or pestered by that pest. However, if the pest difficulty is present in the region and growers have susceptible plants, it is likely they will be beset. Though cleaning fields up may be useful to some degree, it is difficult to fully achieve and thus often ineffective at reaching its objective. Instead, it is often a much better use of time to take steps to enhance crop growth, and leave some appropriate residues in place to support and protect beneficials.

PROTECTING WITH ROW COVERS

Many growers use row covering materials to protect crops from insect damage. These materials often enhance growth as well by shading the crop, holding in moisture, and diminishing wind. Their use also can increase disease potential if conditions under the covers lead to an imbalanced state of growth. It's important to understand the effects of different materials and grades of row cover. Often crops are covered only in early stages of growth when most susceptible to damage, and then uncovered when the crop has grown large or pollination is needed. It is best to wait for a cloudy mild period to uncover crops, so as not to shock the plants with an extreme change. Covers are a very useful tool for growers to avoid insect damage, but as soils develop and the grower's understanding of the environment increases, covers generally become less necessary, and may eventually be unneeded for insect protection.

Growers can engage in handpicking insect pests or diseased leaf parts; occasionally this is worthwhile, as with an aggressive tomato hornworm or a few stray overwintered potato beetles. In general, however, all the control options described above will make handpicking unnecessary, as it is usually a slow and unprofitable task. Time may well be better spent on improving crop growth conditions.

CHAPTER 13

Producing Vegetables Year-Round

Year-round vegetable growing in a cold climate can be challenging, but there is potential for significant profitability. Cold temperatures and low light conditions can be moderated through various structures and techniques. However, to be successful in this endeavor requires an even more thorough understanding of the growing environment than developed through outdoor growing in warm weather. The development of strong, balanced soil fertility is primary in assisting crops to survive the added challenges of off-season growing, which can easily ruin crops.

At Tobacco Road Farm we work with protective structures on an acre (0.4 ha) or more of growing area. This allows us to sell vegetables into a very hungry market at times of year when prices are potentially better, and profitability potential is high. The consistency of year-round production and availability is very attractive to buyers, many of whom hold this as a primary factor in their purchasing decisions. It also allows us to maximize output from the growing area, thus reducing the total area we need to farm, which keeps costs down. As well, the workload of production is spread out over the entire season, giving everyone plenty to do throughout the year.

The two main mechanisms by which plants resist the damages of freezing temperatures are to break down starches into simple sugars, thereby creating high-sugar sap, and to desiccate themselves. Both of these modifications lessen the ability of water to freeze in the plant tissue and thus cause damage. Growers can assist crops by creating conditions favorable to sugar and starch accumulation and by limiting potential moisture exposure. Sugar and starch accumulation is the result of healthful growing conditions, and thus we employ all the factors described in this book that contribute to crop vitality. In addition, protecting crops under structures can help regulate atmospheric moisture.

A highly functioning soil leads to better freeze resistance for crops, and the soils themselves are better able to resist freezing due to the heat generated by the soil life and the aggregation of soil structure. This moderation of soil temperature due to biological activity is critical for successful off-season crop growth. As well, the common darkening of the soil color as fertility develops is better able to capture the sun's warmth than a lighter-colored soil. When soils are functioning to this level, winter vegetable production becomes much more possible. Trying to grow winter vegetables on weak soils is particularly challenging.

Vegetables capable of saturating their saps with an additional sugar lode improve greatly in flavor upon the onset of frost or freezing conditions. Crops that are unable to do this simply die when temperatures hit the freezing point. The increase in sugar content and improvement in flavor at the onset of cold weather does not go unnoticed by the customer! This is a time when vegetables grown in cold climates can vastly differentiate themselves from their warm-region counterparts. Carrot, cabbage, parsnip, leafy greens, brussels sprout, turnip, and rutabaga are a few well-known examples of crops that sweeten up in cold weather. It is generally true that the colder it gets, the sweeter the vegetable, but beyond a certain cold threshold plants will suffer damage. This damage may be temporary, such as water-soaked–looking areas in leaves that may develop following a frost. In vigorous crops these areas rapidly recover once temperatures rise above freezing, but at some point damage can become permanent. The temperature that leads to permanent damage varies among crop species and varieties, as well as the degree of plant vitality. This is another reason that careful soil and environment management, as well as seed breeding, are priorities for growers who embark upon year-round production.

Planning the layout of the cold-season vegetable production area is of even greater significance than designing field layout for the warm season. In the cold season the sun moves in a low arc across the southern sky, which may result in shading difficulties not encountered during the warm season. The cold season is also often a wetter season, creating more challenges in managing soil water. With the lack of foliage on trees and loss of their wind-moderating influence, cold winds can also lead to damage.

Full southern exposure and a slope toward the south can be particularly useful in collecting sunlight during the cold season. Fields thus situated are by far the best positioned for year-round growing. Drainage management is crucial due to increased moisture in the cold season, which is potentially damaging to crops. Raised beds, arranged appropriately if on a slope, and soil aggregate development are primary to optimize drainage. Subsurface tile drainage in the field or ditches around the growing area may also be beneficial because groundwater levels in winter can be high. The development of appropriate windbreaks, as described in chapter 1, for cold-season growing is also of a high level of importance.

In the last few decades, new types of structures have largely replaced glass greenhouses and cold frames. Clear plastic polyethylene sheeting is frequently used for covering greenhouses. Low tunnels covered with this sheeting have largely replaced cold frames. Polypropylene cloth row covers have supplanted the woven straw rolls and mulches of the past. These modern materials are popular because they are very efficient and are sold at a relatively low cost. Unfortunately, they are relatively toxic to manufacture and dispose of and are capable of leaching plastic compounds into soils and plants. It appears they are more capable of off-gassing and leaching when new, much less so when aged. This movement of plastic compounds into the growing environment does not appear to be excessive and the functioning soil environment is likely to decontaminate in a timely manner, but still, less is probably best.

The benefits of polyethylene and polypropylene sheeting are extensive. With proper application the materials can increase solar heat gain, retain soil warmth, and decrease or eliminate cooling winds. They also serve as a barrier to excessive moisture and thus create drier growing environments. Clear polyethylene sheeting is superior to polypropylene row cover in terms of capturing solar heat, eliminating cooling winds, and blocking moisture. Row cover is superior for retaining warmth for nighttime protection, and can be utilized without supporting hoops, which often makes it easier to work with in a field setting. Depending on which conditions are sought, the materials can be used in sequence, or even in combination.

Working with Row Cover

Cloth row cover for cool-season vegetable growing is often placed upon crops during their harvest period in the fall. The material acts to retain heat from the still relatively warm earth, and this protects the crop. The material can be placed directly on the crop instead of supported on hoops and may be double- or triple-layered to provide the necessary temperature to forestall damage. The row cover does block light and retain moisture as well as diminish wind, so this may increase disease and insect pressure if growing conditions are already imbalanced. However, this is not usually a problem if growing conditions are well balanced or otherwise prepared for this alteration of the growing environment by grower action to compensate for this movement toward crop lushness.

In the fall we place cloth covers over crops before cold nights, then remove the covers when weather improves. This can be done relatively quickly as long as the covers are not overly secured. We use the same type of sandbags to secure row covers as we do to hold down solarization covers: UV treated 6-mil sandbags, placed about every 20 feet (6 m) down the row. If we expect to be covering and uncovering the crop frequently, we secure one side of the row cover fully, but on the other side, we place only as many sandbags as needed to secure the cover against the wind and cold conditions expected that evening and overnight. We frequently cover two or more adjacent beds, and we overlap neighboring covers and have sandbags do double duty for those. A cover 150 feet (45 m) long is the maximum length that two people, one at each end, can quickly pull over crops; shorter lengths are certainly easier to handle. Unless faced with heavy wind the two people can move down the field, one person at each end of the 150-foot lengths, securing only the ends with sandbags. They then go back and place the middle bags as needed.

For cold protection in the fall, we often start by covering a bed with a single medium-weight row cover at first frost and then progressively add more covers on top as the temperatures decline. We may protect some hot-season crops like pepper or eggplant through one or two fall frosts, but once frost has started to hit, overall conditions are too cold for these crops to continue to thrive. We use row covers extensively to protect fall root vegetables and leafy greens that are cold-tolerant but can be damaged if temperatures drop too low. These include cabbage, turnip, kale, beet, carrot, broccoli, parsley, radish, cilantro, and celery. Such protection greatly increases their harvest period. As well, the late final harvests from the field are the sweetest, and the crops keep well for an extended period when finally brought into root cellars for winter storage and sale.

We generally work with 10-foot (3 m) wide row covers for both 36-inch (1 m) and 58-inch (1.5 m) bed widths. We cut row covers to 50 foot (15 m) or 150-foot (45 m) lengths depending upon field layout. Standard sizing of lengths and widths is important for efficiency of reuse. The covers last three or four seasons, possibly longer, but damage by deer can shorten their life span. Deer are particularly hungry during the fall season when these covers are in use, and deer hooves are very effective at punching holes through polypropylene. If the holes are not too severe, we keep the covers in service, because when the covers are placed over crops in multiple layers, the holes will not align. Deer generally do not damage crops covered with clear plastic sheeting, so sometimes we add a layer over the top of the row cover fabric as a preventive measure. When not in use the covers are dried out, rolled up, and stored in sheds that are—hopefully—rodent-proof. When laying out or rolling up covers, we are careful to avoid breathing the dusts that puff up off them.

Row covers are prone to freezing to one another and to the ground, which can make it difficult to avoid damaging them when moving them aside while harvesting. To avoid this dilemma, we try to remove the row covers from the field before late November, the time when snow cover sets in for winter. (Well, at least that's when snow cover used to set in for winter.) If there are still harvestable crops in the field, we may cover them with clear plastic sheeting instead; it does not readily freeze to the ground. For additional insulation, straw can be placed upon the crop underneath the sheeting. This is useful for winter harvesting of carrot and parsnip.

Working with Low Tunnels

In the spring polypropylene cloth row covers work well for protecting warm-season crops from potentially damaging nighttime temperatures. These crops are often transplants set from out of the greenhouse. We suspend the covers on hoops to minimize abrasion of tender seedlings. This use of low tunnels also creates a somewhat protected environment that helps harden off greenhouse-grown transplants before they are fully exposed to the outdoor environment.

As described earlier in the book, the standard low tunnel on our farm covers the 36-inch (1 m) wide beds and is 40 feet (12 m) long. This relatively narrow profile allows for greater resistance to snow loads, which can potentially collapse low tunnels. The 8-inch (20 cm) wheel track between beds provides space to secure the edges of covers on the two bordering beds. The bed length of choice for covering with plastic (polyethylene) row covers is only 40 feet, compared with the 150-foot-long polypropylene covers. At a 40-foot length, two people can easily pull the plastic covers over the hoops as they work their way down the field. This short length is particularly important when the covers are laden with water and snow in winter, which makes them much heavier. Shorter covers are also less at risk of being loosened and blown off by wind.

MAKING HOOPS

The thinnest commercially available wire hoops for low tunnels are ⅛-inch (3 mm) diameter; ³⁄₁₆ inch (5 mm) is somewhat common. We prefer ¼-inch (6 mm) diameter, however. Some growers make hoops out of thin metal pipe, like electrical conduit. The choice of hoop material basically depends upon the grower's environment. The thinner-diameter hoops are useful for conditions without snow, and of course they are less costly. Hoops with a ³⁄₁₆-inch-diameter can hold up fairly well under snow load, especially if they are closely placed. Thicker wire hoops and pipe hoops fare the best at resisting snow loads. Since our region sometimes experiences snowfall of 1 to 2 feet (30–60 cm) during a single storm, we use ¼-inch solid steel hoops. The solid nature of these hoops makes them easier to drive into the ground than hoops made out of metal pipe; they are less expensive than pipe as well.

FIGURE 13.1. Beds laid out for low tunnel assembly. Bags first, then hoops, then covers. This picture dates from our tillage days. Straight raised beds with exposed soil surface, mulched wheel tracks to keep weeds out and prevent bags and covers from freezing to the ground. The field does not look like this anymore.

Since precut, bent hoops of ¼-inch-diameter steel are not available on the market, we purchase ¼-inch-diameter round straight stock through metalworking shops and make our own. The stock comes in 20-foot (6 m) lengths. We have each 20-foot length cut or sheared at the shop into three equal pieces (6 feet, 8 inches / 2 m each), an appropriate size for 36-inch-wide beds. We do make some longer hoops for special purposes as well. The steel hoops are not galvanized. We prefer that because galvanized coatings may contain metals that we do not necessarily want to add to our soils. (Most commercially available precut hoops are galvanized.) Steel hoops rust faster than galvanized hoops, so they may have a shorter life span, but still, they provide decades of use. The rusted surface of the hoops helps the cloth row cover better resist the wind, because the cloth adheres to the rust somewhat. The round straight stock is coated in an oil, so we do give it a wash with liquid soap and warm water from the hose before we work with it. We bend the cut lengths of steel into a U-shape by hand. Hoops sometimes become distorted during seasonal use and require a little rebending. It is important to maintain the U-shape, because a consistently arched hoop tunnel with straight sides driven into the earth better resists collapsing under a load of snow.

FITTING MATERIAL TO HOOPS

Ten-foot (3 m) wide cloth row covers fit over 36-inch (1 m) wide hoops with plenty to spare on the sides for holding down with the sandbags. For many years we used covers that were 8 feet, 4 inches (2.5 m) wide and that was sufficient, but we standardized to 10-foot-wide because that width works well for covering both the hoops and tall crops (without hoops) growing in the 58-inch (1.5 m) beds. For plastic sheeting, we utilize 3-mil greenhouse clear plastic for the outer layer and 2-mil perforated clear plastic for the inner layer. Thus we cover each hoop with two layers of plastic. We purchase unperforated 3-mil greenhouse plastic in 24-foot (7 m) wide sheets. To cut the plastic into 8-foot (2.5 m) widths, we pass a long pipe through the cylinder at the center of the sheeting and then suspend the roll and pipe upon two sawhorses. We hammer a couple of nails into the sawhorse on each side of the pipe to prevent the roll from falling off. The plastic is measured to locate the 8-foot mark among the folds. Two people hold small sharp knives at these points, cutting the plastic as another person pulls one end of the sheeting down the field. Upon reaching a 50-foot (15 m) length, the plastic is cut off horizontally from the roll and the process begun again. When cutting plastic, it is beneficial to maintain steady tension so that the knife can just slide along. It takes practice to gain proficiency at this, and to avoid frustration it's helpful to have an extra person on hand to hold the roll taut while others cut. All low tunnel covers are cut to 50 feet. This provides for covering 40-foot (12 m) beds with enough extra to cover the endwalls and to extend a few feet farther along the ground surface, providing a lip of plastic that can be secured by sandbags when needed. This greenhouse plastic lasts about four seasons of use.

The undercover of 2-mil plastic comes thoroughly perforated and allows for excellent ventilation. This material is purchased in 33-foot (10 m) wide rolls, which fortunately are folded so that four 8-foot, 3-inch (2.5 m) pieces result if the three seams at one edge of the roll are cut through. The sawhorse method works for these rolls, too, or they can be rolled out on the ground and cut. Sometimes the manufacturer's hole-punching equipment does not fully dislodge its "chad" during the perforation process, and we are faced with the dreaded hanging chads. If this is the case, after cutting the roll into pieces, we manually collect up the chads. This is a tedious process, but we do not want chads to end up in the soil (at least we don't have to count them). Perforated covers commonly last three to four seasons, after which they become brittle and start to rip easily.

SETTING HOOPS IN THE FIELD

The first step in the process of setting up hoop tunnels is to place the sandbags in proximity alongside the beds. This allows us to avoid the awkward task of carrying sandbags down the wheel tracks when the hoops are already in place. The sandbags are filled about two-thirds full with sand or subsoil and weigh about 20 pounds (9 kg) apiece. We keep the sandbags alongside the growing area in a shady spot when not in use, so there is usually a supply readily available to whatever area we plan to cover. To move them out into the field, we use a tractor, truck, or simply wheelbarrows.

We place five bags along each side of a 40-foot (12 m) tunnel, one at each end, and three down the row at 10-foot (3 m) intervals. More may be required in windy locations, less if we're covering hoops with cloth row cover instead of clear plastic. The bags last many years and mostly fail due to abuse from being thrown around, ripped up when frozen to the ground, walked on, driven over, and so on.

Hoops are then set. Spacing of hoops depends on the conditions: 2 feet (60 cm) apart for good snow load resistance; 4 feet (1.2 m) if no snow load is anticipated. One person holds several hoops in one hand while walking backward along a wheel track. A second person walks along the other side of the bed. The first person flips the end of hoop across the bed to their partner, who sets the hoop at the proper spacing and drives their end into the ground three or so inches. The person holding the hoops then drives their end in, takes a step or two backward, and flips another hoop end over to their partner. The work proceeds until the bedding areas are fully covered. This can be done quite speedily if the partners are proficient and focused.

For plastic-covered low tunnels, once the hoops are set the perforated covers are stretched out along the side of the area to be covered. This also requires two people, one at each end of the 40-foot bed sections. We pick up a cover, move it down the field, and stretch it over the hoops covering a bed. This also works best when the partners are attuned to each other, and it is difficult to communicate by talking while placing plastic because of its rustling nature. The cover must be placed evenly side-to-side and along the full length. The partners apply tension along one side, place the end sandbags, then bring tension to the other side and place the next end sandbag. We repeat this process until we have placed all the perforated covers needed on that day. Then we cover each perforated cover with a solid plastic cover, using the same technique. Once the solid covers are all in place, the middle bags down the bed are then set since the bags are now underneath the plastic. This requires pulling the bags up between the two layers of interlaying plastic and then using our feet to make sure the plastics are again laying fully across the wheel track, and placing the bag back down upon the overlapping plastic. This technique works well for placing and securing cloth row covers over hoops, too.

We start to set up clear plastic low tunnels beginning in October when frost becomes a threat. At first, we might set only the perforated plastic cover; it limits wind and excess moisture while still allowing most of the sun's light energy to reach the crop. When additional heat is required, we add the solid plastic layer. The double cover results in about a 35°F (19°C) temperature gain on a sunny day. So if it is 30°F (–1°C) outside, it's a windless 65°F (18°C) with appropriate moisture under the covers. Overheating under the covers is a particular danger in early fall. If the outside temperature reaches 60°F (15°C), then it is 95°F (35°C) or more under the cover and venting is required. The two-layer cover is a great advantage here, because we can pull back the outer cover, leaving only the inner cover in place. Crops growing in such a warm, humid, windless manner cannot be suddenly exposed to outside conditions without being damaged. The perforated cover provides an intermediate condition between fully covered and uncovered.

MANAGING COVERS IN DIFFERENT CONDITIONS

The combined use of solid and perforated plastic covers can give excellent interaction with environmental forces. They allow relatively quick exposure to full rain if both are removed, or partial rain if the perforated cover is left in place. We can also completely keep rain off the crop by leaving the solid cover in place. The same management options are possible with regard to wind. The covers allow excellent light and warmth accumulation on sunny days with nighttime temperature elevated as well. If even higher nighttime temperatures are needed, cloth row cover can be placed directly upon the crop under the low tunnel, and there will be no danger of the cloth cover freezing to the ground. Cloth covers can also be temporarily placed on top of the plastic-covered low tunnels to raise nighttime temperature. Watchfulness is important so they can be removed before freezing conditions would consistently freeze them to each other or the ground. Cloth covers can also be utilized in a similar manner to the inner perforated cover, though freezing to the ground and light limitation may present difficulties. The superior ability of low tunnels to allow managed interaction with environmental forces enables a grower to create conditions of balance and enhance crop quality.

Atmospheric water content has an influence on the balance of growth. In the tunnel environment the air develops a high moisture content, which can be regulated with the introduction of outside air. Appropriate venting of tunnels is particularly important for cool-season crops. Sometimes a humid environment is useful for growth and sizing, but it can be overdone and cause weakness in the crop. To harden a crop by introducing outside air, the solid outer cover on a low tunnel can be partially or fully pulled back. This venting also reduces the tunnel temperature, of course. We often vent tunnels frequently in the fall, much less (or not at all) during the colder part of winter, then again more frequently as conditions warm up and growth restarts in the late winter and

FIGURE 13.2. The ends of the low tunnels vented under mild winter conditions.

spring. To partially vent a low tunnel, the solid cover is pulled out from under one or more sandbags on one side and secured under the end bag of the other side. For full venting the entire cover is released by removing all the sandbags along one side, pulling the plastic over the hoops from end to end, and laying it between the beds.

This is generally the manner in which low tunnels are worked: Every other wheel track has its covers pulled off when venting or fully uncovering and placed on the opposite side. This gives the field an "every other" open wheel track where we can freely walk and work the beds that are uncovered. We consistently follow this pattern when covering, venting, and uncovering. Thus the sandbags along one side of every tunnel are left in place and never moved from the time of initial setup until we take down the tunnels in spring. When re-covering, the side of the tunnel that was uncovered has the appropriate number of sandbags replaced according to environmental conditions. In mild times this often may be only the end bag with no middle bags placed. This allows for very swift uncovering, venting, and re-covering, with two people able to rapidly adjust the covers as they move down the beds. The covering of larger areas with low tunnels adjacent to each other with overlapping covers in the wheel tracks creates growing "blocks" that are better able to hold and harness

FIGURE 13.3. Tunnels under full snow coverage are well insulated against cold temperatures but do not allow in enough light for much growth.

solar energy. This is superior to single isolated tunnels where the soil alongside each tunnel is exposed to atmospheric cooling and other conditions. Setting up of blocks of low tunnels also allows us to control the flow of excess moisture from the greater area, depending on the site slope and tunnel layout. Blocks of low tunnels also protect one another from wind much better than stand-alone tunnels. Well-placed windbreaks along the windward side of these blocks can be very helpful as well. Snow upon low tunnels is both of great benefit and a possible drawback. Snow of less than a foot in depth generally slides off the covers on its own into the wheel tracks. This snow then holds down the covers, which further insulates the adjacent growing areas and prevents wind from pulling off the covers. The snow also reflects sunlight into the growing area, improving growth. Sometimes the snow is deep enough to cover the tunnels. This can offer great insulation and protection from cold outside temperatures. Snow cover is therefore very beneficial before a cold snap. Once sunlight can penetrate the tunnel to some degree, the tunnel starts to heat and the snow will slide off and begin melting.

The disadvantage of snow is that it can make harvesting more labor-intensive. The covers have to be pulled out from under the snow, then placed on top of the snow in the wheel track, and sandbags have to be pulled up and

221

FIGURE 13.4. Even under deep snow, covers can be pulled out so harvesting can proceed if needed.

placed on top of the cover. This can be particularly difficult if the snow is deep. Another difficulty is very large volumes of snow. Snow accumulations of 2 feet (60 cm) or more can completely bury the low tunnels, which then allows the crop very little light and thus can slow growth. The crops are well protected in this environment, however, because the ground is usually not frozen under low tunnels and will remain that way under heavy snow. Essentially the crops just stabilize and wait.

If we want to expose crops to light, we rely on a simple method that employs plastic snow shovels whose blade corners have been trimmed round and smooth so they won't pierce the plastic covers. We trudge down the snow-filled wheel tracks, using the shovels to pull a little snow off the tunnel tops and into the track. A little at the top is enough. Once the sun can penetrate the tunnel, the air inside will begin to heat and further melt the snow.

In the occasional winter when snow depth builds up to 3 or 4 feet (1–1.2 m), we take a little more aggressive action. We wait for a day when temperatures rise above freezing, and then we fill several 55-gallon (210 L) barrels at the field edge with water. We then lay down some old plywood sheets near the barrels, dump a pile of very dry sawdust on top of the plywood, and toss a match on it. We get our sawdust from a coffin maker, but any dry sawdust from cabinetmakers

FIGURE 13.5. We use a modified snow shovel to expose the top of the tunnel, which lets in light and thus begins melting the snow.

or such will do. The sawdust readily burns without much flame. We stir it with long-handled rakes for 20 to 30 minutes. Once the sawdust has consistently charred but has not yet turned to ash, we scoop it up with flat-bottomed metal shovels and toss it into the barrels of water. To these barrels we also add about 100 pounds (112 kg) of sea salt and 10 gallons (93.5 L) of molasses per acre (ha) we plan to "desnow." Next we hook up the irrigation pump, using a short suction hose that has a mesh screen across the end. We use a long hose 1 inch (2.5 cm) in diameter as the discharge. One person stirs the salty, sugary sawdust and water in the barrel and makes sure this slurry is feeding through the screen into the pump. Another person maneuvers the hose around the field and sprays this material on top of the snow. There is a fair bit of pressure, and the spray reaches about 60 feet (18 m) across the field. This is an impressive undertaking well worth observing.

This material is very effective at melting snow. The char melts snow, the salt melts snow, the sugars melt snow, and the sun shining on the black char melts snow. We have seen feet of snow disappear after a few days. Another benefit of this is that the materials we use are all of benefit to the soil. The covers protect the crops from becoming sooty. The soot eventually washes off the covers into the wheel tracks, but care must be taken to shake off the soot to one side when opening the cover.

FIGURE 13.6. Light snow is helping to hold the covers of these low tunnels in place.

As temperatures warm up in the spring, we remove the covers in stages. First, the solid cover is more and more frequently left off. Eventually the perforated covers can also be removed. When it is time to remove the perforated cover altogether and expose the crops to the outdoor environment for the rest of the growing season, we watch the weather forecast. It's best to make this change when conditions are cloudy so as to more gently introduce the plants to the outdoor environment. We pull the covers out of the field and lay them out to dry to a fair degree. They are then rolled up and stored. It is particularly important to expeditiously remove the plastic covers from the field. If left sitting in the wheel tracks in warm conditions, they pool water, which attracts and drowns earthworms. This quickly begins to smell very bad!

We prefer low tunnels to greenhouses because of the inexpensive nature of the low tunnel and because they allow us to more closely interact with and respond to natural forces as our crops grow. Initially, the ability to employ a system of mechanized field culture and then shift into winter crop protection mode and back to mechanized field culture was very attractive to us. Over time, however, it has been the ability to better maintain soil fertility that has led us to stick with low tunnels rather than switching to the high tunnels that so many growers employ. The superior development of soils under low tunnels results

FIGURE 13.7. We prepare a sawdust char to melt snow, and the results are rapid, even when the snow covering is heavy.

FIGURE 13.8. An example of spring row covers. Spring-set transplants are under the cloth row cover (*left*); overwintered lettuces have the solid plastic fully off, leaving only the perforated cover in place (*right*).

from the fact that soils are less cut off from atmospheric conditions than they would be in high tunnels. With a low tunnel growing system, there is a shorter period of coverage when compared with high tunnels. As well, the soils under low tunnels stay more naturally hydrated through the wicking of moisture that penetrates the ground in the wheel tracks. Thus the tunnel environment maintains a much more natural balance of soil water to air.

High Tunnels and Greenhouses

High tunnels and greenhouses have a long history of soil difficulties. Growers deal with this in different ways including movable units, taking the plastic covering off the tunnels for periods of time, replacing the soil in the greenhouse/tunnel area, or not growing in the soil at all. Probably the most common approach is to remove the plastic covering from the high tunnel for a period of time. Often growers who cultivate hot-season crops in high tunnels remove the cover in winter, and the growers of cold-season crops pull off the covers during the warm season. In some old glass greenhouses, the soils were traditionally replaced every 10 years, sometimes on a schedule of $\frac{1}{10}$ replacement per year. Other greenhouse and hoophouse growers have shifted away from growing in the ground altogether, and grow crops in containers instead.

All of this is an effort to make up for the difficulty that these covered structures present to soils. It is likely that the cause of this difficulty lies in the ability of structures to cut off the soil environment from the greater environment. This ability to be isolated from environmental conditions can be potentially rewarding for growers yet can also offer the challenges of a disturbed soil system. In addition to the above-mentioned techniques, growers can manage the greenhouse environment to the best of their abilities in order to make the soil as balanced as possible, nutrient-rich, and full of life. This can supplant, to some degree, the more drastic techniques mentioned above. Some helpful management techniques include no-till, careful nutrient additions, covering the soil with mulch and growing plants, maintaining ventilation, and keeping temperatures from reaching extreme highs. Irrigation is of central importance to maintain a balanced soil air:water ratio at all times. Excessive drying and lack of steady moisture in covered soils is a likely cause of soil biological dysfunction and contributes greatly to the buildup of salts. Of course in this condition irrigation water quality is of increased importance.

Large growing structures offer some attractive advantages for growers compared with low tunnels. They are conducive to artificial heating, seedling production in potting soil media, and growing tall crops. They allow easy access to crops in winter. Low tunnels can be set up inside these structures to create a double tunnel system. This can greatly assist crops in winter hardiness.

FIGURE 13.9. Low tunnels opened up for rain; the high tunnel visible in the background cannot be uncovered as easily.

On our farm we utilize two high tunnels for the winter growing of various greens, especially for taller crops such as kale, as well as seedbed production of large volumes of alliums including onion, leek, and shallot. Often we grow salad green lettuces in one of them and apply a second covering with a low tunnel structure. We wait to cover the high tunnels as late as possible in the fall, potentially using low tunnels or direct row covering first then adding the larger tunnel cover as temperatures diminish. In the spring we uncover the tunnels as soon as possible, again potentially with the assistance of low tunnel or row cover. During the warm season, vegetable crops or cover crops are seeded into the uncovered tunnel area.

A third hoop tunnel remains permanently covered. In this tunnel we never grow crops directly in the soil. Instead we use it as a seedling house to grow transplants in flats in spring. After seedling production we place shade cloth over the structure and begin drying seed and crops on shelves built up inside. For the duration of these drying projects, the hoop tunnel is kept as dry as possible, watering the soil surface only to keep down dust if necessary. First we dry the overwintered seed crops like mâche, claytonia, brassicas, and lettuces. Then we bring in the garlic harvest to dry, followed by the shallot and onion harvest and more seed crops. In the fall we remove the shade cloth and cure our winter squash on the shelves for a week or two. After this the tunnel is converted to a winter vegetable washing area.

As an exercise, I added up the costs of materials and divided the cost of covers and sandbags by 3 (a conservative estimate of their life span) and the cost of hoops by 20. This revealed our total cost per acre (0.4 ha) per year to cover our crops: $4,014. The return we can generate by growing crops year-round rather than only during the regular growing season far exceeds that figure.

CHAPTER 14

Harvest and Marketing

Harvesting and packing produce for market are two of the most labor-intensive aspects of growing vegetables. Developing an efficient system is imperative for success. All of the grower's effort to raise a crop that is abundant, weed-free, and undamaged by insects and disease add to harvest efficiency, and such healthful crops have a long shelf life after harvest. Thus all the careful soil and environment culture techniques described thus far in this book have a direct relationship to harvest efficiency and the farm's profitability.

The two areas of grower effort most directly related to profitability are putting in a crop and harvesting the crop out of the field. All other aspects of grower effort are essentially to make sure that these two events can occur efficiently. As the soil and the grower's abilities improve, more labor can be focused on these two areas. This chapter focuses on efficiently getting the crop out of the field and into the marketplace.

The Flow of Harvest Work

Some types of vegetable crops maintain best post-harvest quality when picked during the cooler times of the day, usually morning or evening. Harvesting generally begins in the morning with leafy greens—the crops that need to be kept most hydrated—and progresses to those for which there is less concern for hydration, such as tomatoes and other summer fruits, followed by full sun harvest of root crops like potatoes. Some crops, such as summer squash and zucchini, need harvesting every other day over a period of time; other crops, like garlic, are harvested all at once. It is important that humans engaged in harvesting position themselves well for efficiency both in terms of how and where they are harvesting and how they maneuver their body during harvest. For instance if harvesting down a row or bed,

FIGURE 14.1. Harvest of salad greens. Note the elbow on the knee as the accumulated greens in hand are tossed into an appropriately placed basket. The next cut is already being decided upon as this proceeds.

the knife hand is closer to the side of the bed, and the holding/bunching hand is closer to the wheel track where the harvest bushel is placed. Harvest containers, carts, trucks, and various equipment need to be fully planned out for efficiency. The harvester needs to understand the size, ripeness, and part of the plant that is being harvested, and more. The body needs to be properly positioned for maximum speed as well. Often this is in a bent-over position with one elbow on one knee. The arm that is on the knee has the hand that collects the leaves, fruits, or roots. The other hand holds the knife or does the picking. With knife harvest the support elbow can switch back and forth as harvest progresses. Thus the back is supported, which allows the harvester to continue for extended periods.

Once in position, it is the rate of decision making that determines speed. The harvester's ability to quickly identify the portion of a crop to be harvested is critical. This may require some training. For harvest to proceed smoothly and efficiently the harvester identifies the proper portion to be harvested, then while the fingers are actually doing the harvesting the eyes are looking ahead for the next portion to harvest. If knife-harvesting, the harvested portion is consolidated in the non-knife hand as the knife hand is moving forward toward the next cut. If hand-harvesting, the process is similar, with the transfer to the holding hand taking place as the next portion to be harvested is identified.

FIGURE 14.2. Cilantro, dill, and basil are harvested with about 1 inch (2.5 cm) of root attached, using a serrated knife.

When it is considered how fast a knife can swing or how fast a hand can snap off a portion of crop, it becomes clear that the rate of harvest is determined by the speed of decision making, not by the ability to physically swing a knife. Harvesting is often one of the most labor-oriented efforts on the farm; it is very profitable to have this be done speedily and accurately. It is also excellent at developing the skill of decision making, which benefits all other areas of production as well.

The harvester needs to quickly and repeatedly decide what and how much of a plant is appropriate to cut or pick. When harvesting leaves, the goal is neither to overharvest, which delays crop regrowth due to lack of leaves (and thus photosynthetic capacity), nor to underharvest. Market demand is another factor to consider, though, because if a set quantity of a crop is needed that day, it must be harvested. In this regard harvest can proceed much more efficiently if there is always a plentiful supply, so planting in excess of projected demand can lead to significantly increased harvesting efficiency. On the other hand excessive crops that are unsold bring down profitability because the effort put into growing them is largely wasted. This is particularly true of crops that need constant harvest such as beans, cucumbers, and summer squash. With these crops, the grower can be forced into harvesting fruits simply to keep up overall yield and promote quality of the successive fruits, even if there is no market for that day's harvest. Overall, a bit of excess in planting, but not too much, is useful to buffer against potential crop losses and promote harvest efficiency.

When harvesting whole leaves from plants such as kale, chard, or collards we try to harvest as much of the stem as possible, because stem stubs left behind often rot on the plants. When using a cut-and-come-again harvest technique for crops like salad greens or spinach, we cut all the large leaves at a relatively high level in order to avoid damaging the small inner leaves; this enables fast regrowth. Some crops, including cilantro and basil, are cut just below the soil line and bunched. The stubs of roots on the harvested stems help maintain post-harvest quality, particularly of basil, which is intolerant of refrigeration, and is kept with the stem ends immersed in water.

Harvesting whole plant parts, such as many of the root crops, lettuce heads, celery, and cabbage, is more straightforward but still requires the speed and efficiency of focus and decision-making skills. The improvement in crop and soil quality from no-till in particular has made for improved ease of harvest of root crops, which is a great savings of labor hours. The pulling of garlic and carrots can be quickly performed without the use of digging tools. Even parsnips can often be thus uprooted. All the root vegetables come out of the ground much more cleanly due to the highly aggregated condition of the soil.

Harvesting Equipment

We keep a variety of sizes of harvesting knives on hand. Straight-edged knives for harvesting leafy crops need to be kept very sharp by frequent, almost daily sharpening (a couple of quick passes on a grinding wheel). Serrated knives need less frequent attention, but sharpening them requires more effort. We secure a serrated knife in a vise and use a small round file, usually a ⅛-inch (3 mm) round chain saw file, to improve the serrations. Serrated knives are used to harvest hard-stemmed crops such as winter squash, kohlrabi, eggplant, and okra as well as to undercut crop roots like cilantro and basil. They also work well for various fruit harvests including pepper, summer squash, and melon. We prefer to use hand pruners to harvest strong-stemmed tomatoes.

High-quality wooden bushel baskets serve well for gathering many types of crops. Low-quality baskets have weak, excessively folded bottoms that break easily. Natural wood baskets are a fine choice for vegetable harvest because they are a generous size and are made of non-toxic, breathable material. The baskets are an appropriate height to gently lean upon while harvesting, thus assisting with back support. Four baskets are generally carried at once by linking two wire handles together in each hand.

For harvest of root crops, we utilize boxes made on-farm. These boxes are made with standard-size bottom dimensions so they can be stacked, and of varied

FIGURE 14.3. Potatoes lift easily from our soils, and we transfer them to harvest boxes like this one, which holds a half bushel. This box has an aluminum screen and hardware cloth bottom, but we now make most boxes with a wooden slat bottom.

heights to accommodate from half a bushel to a full bushel (17–35 L) volume (see figure 14.3). The boxes have slatted bottoms and runners for feet, which allow some air to circulate when stacked. Separate wooden lids are available if needed. These boxes are very sturdy and easy to repair. They last for many years.

Processing the Harvest

When we cut leafy greens on a sunny day, if harvest is proceeding slowly, we move full baskets into the shade or cover them until harvest is complete. When we bring the baskets in from the field, we place them in a shady area by the processing area. A hose is used to mist their surfaces and then they are covered with canvas cloth to await processing. If processing is not going to happen in a relatively timely manner, the baskets are transferred to a nearby root cellar.

When we bring in boxes of roots, we often place them in the root cellar to await processing or, for potatoes, put them on drying racks first. After the potatoes dry, we transfer them to the root cellar. We lay harvested tomatoes in single layers in trays because of their delicacy. We transfer these trays to storage shelves in an open-sided shed, or onto stand-alone covered shelving outside.

We have designed our washing and packing system to flow as efficiently as possible. Picking sheets for the daily harvest need to be written out clearly, harvesting of crops grouped with certain people in certain areas, and with the goal of a steady flow to avoid backups in the chain. The washing/packing area needs to be set up to enhance this flow from field-harvested vegetables all the way through to loading the packed vegetables onto the delivery trucks.

The washing/packing area can generally benefit from a cover to keep off sun and rain as the vegetables are run through. Multiple wash troughs at an appropriate working height, screen tables for drying, and solid wooden tables for cutting and stacking produce boxes all need to be arranged for efficiency and steady flow. A root vegetable barrel washer is an excellent addition to this basic equipment if growing even an acre (0.4 ha) of root vegetables or so. Easy access for delivery trucks, along with proximity to stored supplies, root cellars, and coolers, is also essential.

A clean, uncontaminated source of water is very important. Water from a drilled well is often the best possible source. Municipal ("city") water is often contaminated with antimicrobial chemicals, which will not only contaminate produce with unwanted chemicals but also damage the biology present on the vegetable that offers many benefits including the important benefit of healthful digestion when consumed by the customer. If the only water available is municipal water, it may be worthwhile to consider an investment in water filtration equipment, or having a well dug or drilled.

Root Cellars

Root cellars are extremely useful for vegetable producers. They offer the means of storing and holding large volumes of vegetables without the expense of artificial cooling. The high humidity, cool temperatures, and darkness create ideal storage conditions for many vegetables over the course of the season. In the summer they store the fruits such as cucumber, pepper, melon, and summer squash well. In the fall as the temperatures start to drop, root vegetables that need protection from outside freezing temperatures are steadily brought in. These vegetables are then sold through the winter months to provide winter income and activity. In addition to a large root cellar (20 × 24 feet / 6 × 7 m), we have one small root cellar into which we fit a CoolBot-adjusted air conditioner to provide cold storage during the summer months when required as well. This is particularly useful when tender vegetables are awaiting delivery or pickup. We utilize a cool, dry, mouse-proof outbuilding to keep onion, garlic, shallot, and winter squash from freezing in winter.

Marketing

We pick the majority of crops in the morning and wash, pack, and deliver by afternoon. This keeps the vegetables as fresh as possible and cuts down on refrigeration costs. Most of the crops are wholesaled to grocers and distributors, with farmers markets providing a channel of sale for some of the excesses from overproducing for these wholesalers. We negotiate reasonable prices from the wholesalers and find the wholesale market profitable for us. The higher price received by direct sales can also be profitable, but when we assess the added expenses that goes also with direct marketing, we find a very similar level of profitability. This may be different in other regions or situations but it is definitely an area to assess with an honest, unbiased eye. The market demand for a vegetable largely determines how much of it needs to be grown at a particular time. Markets do vary and are sometimes hard to predict, but again consistency is very important for customer loyalty and satisfaction. In general a little overproduction is needed. Markets can be sought for excesses if this can be done efficiently, or excesses can simply be given away, which of course creates much goodwill. Of course market demand is also related to pricing and the ability of a grower to maintain profitability at the market price. Vegetable growers, like most other farmers, are prone to insufficient market pricing, both in terms of what they ask from the market and what the market asks of them. Farming generally is treated very poorly in the modern system of economic rule. The system's basic tenet is the control and exploitation of natural resources for the gain of those with the most capital. Farmers, crops, and land are all examples of natural resources the system seeks to control. Farmers generally also have very little capital clout, thus they have very little say in how the system exploits them. The result of this is market controls, regulatory burdens, subsidy programs, and other more nefarious activities to the detriment of farm goods pricing in the marketplace. In other words the system is capable of making a tomato from its chosen source of production very inexpensive and also creating conditions to increase the cost of an independent grower's tomato. It is important for growers to understand their position in the system and make appropriate plans to deal with the pressures the system puts upon them. This modern system of rule is extremely pervasive in control of the economy and government and is highly capable of dividing producers and pressuring them to work against one another in the marketplace. The system profits from these conditions as well as many other conditions of human suffering, such as illness, environmental devastation, disasters, and war. Growers are best able to stay profitable amid the overarching economy by working closely together as well as with the customers themselves to keep prices up. By working together instead of against one another, growers can reach higher levels of individual profitability as well as maintain profitability for all.

Growers generally benefit financially from certain crops and less from others. A grower needs to understand the financial ramifications of all that goes into raising a crop as well as marketing it. That said, it may not be the most efficient use of time to try to determine the cost of a crop's production through detailed analysis, but a grower does need to have a sense of what it takes, in terms of labor and materials, to raise a crop and what the market is willing to pay for it in order to help determine how much of it to grow. Some easily quantifiable cost-of-production factors to look at are average yield per area, time required to harvest a crop, and time to process and deliver a crop. Average yield per area is relatively easy to measure; this combined with price paid by the market yields dollars per area. We can quantify this as dollars per square foot. For example, a very easy crop such as winter squash might yield one dollar per square foot, whereas an overwintered multicut salad green under cover might yield five dollars per square foot. Another measurement that is relatively easy to consider is the cost of harvesting, processing, and delivering a crop in terms of human labor. In general we considered it reasonable for this effort to be in the range of 25 to 33 percent of the income generated. When these measurements have been done consistently across crops at different times of the year and through the years, an understanding of the generalities of production costs can be gained. With this, adequate pricing can then be better determined.

Of course adequate pricing cannot always be secured in the marketplace, so production may need to be altered for this consideration. Growers generally need to grow larger amounts of crops that represent higher profitability in order to keep their farms flourishing. However, profitability is certainly not the only reason to grow a crop, because the happiness of growers and customers is a primary concern. Growing an unprofitable crop may bring customer loyalty, which may lead to those customers purchasing other vegetables from you as well. Less profitable crops may also fit in well to the production system or otherwise be of benefit to the soil. The marketplace is a dynamic entity where the best and the worst of human actions are often on display. An honest, loving approach of offering high-quality vegetables is still the means to great rewards.

Thus, in more than one sense of the word, bringing in the harvest is perhaps the most rewarding part of vegetable growing. When the harvest of abundant, nutrient-rich vegetables is proceeding well, a grower can breathe easy in the satisfaction of a job well done and an awareness that many will be well fed.

CHAPTER 15

The Grower

The cultivation of vegetables is, in essence, the application of the human will and desire to an area of land. Nature covers land in perennial plants for maximum stability and enhancement of the living force. Annual plants, with their fast, vigorous, short-term growth, are the plants nature applies to disturbed areas where the earth needs to quickly recover vegetation. Vegetable cultivation requires the removal of the natural perennial vegetative covering and replaces it primarily with the growth of fast-growing annual crops in accordance with the will of the grower. In this sense vegetable growing can be understood as a disturbance of natural systems. With the continued weakening of nature's inherent vitality, it is of ever-increasing importance that the grower is clear in objective, flexible in approach, and careful to work as closely as possible with natural forces in order to make this disturbance a condition of benefit for the living forces of nature. Why do we grow vegetables? At its root, the objective of most growers is related to a basic human need for health and happiness, though for some this may not be the case! Most who pursue vegetable farming and gardening have a primary concern of providing healthful foods for themselves, their family, and the greater community, the process of which can lead to a well-lived, largely independent harmonious life with greater potential for happiness than many other pursuits of livelihood. Profitability often stems from this basic undertaking, thereby providing the grower with a consistent livelihood. It is worth considering, however, that if profit is an individual's main goal, it may be wiser to engage in any number of other careers more profitable than vegetable farming in America.

Though deeper spiritual goals may well be the motivation behind the desire for health and happiness and their pursuit through farming, these concerns are still well worth considering as primary objectives of agricultural undertakings. To bring health and happiness to self and family is certainly a worthwhile goal. However, when we expand this objective to the greater community, even

237

greater reward is possible. A life of devotion to providing for the basic needs of others brings much to the individual, not just in terms of spiritual rewards but also to the more tangible concept of profitability. To a significant extent the greater community of people in the United States is dependent upon others to raise food for them, and their gratitude is profound when high-quality foods are provided. This gratitude produces all kinds of effects such as customer loyalty, increased effort from farmhands, and people bending over backward to be of assistance. A simple example of this is the trucker who loads and delivers wood chips for composting at Tobacco Road Farm. Since he understands the agricultural undertaking that we are engaged in, he absolutely overstuffs his trailer with wood chips to give us as much value per load as possible. This is the general rule in many of our dealings with our suppliers and customers. Their wish is to give back to those who give to them and the greater community. That trucker doesn't even eat our vegetables; he raises his own, yet he understands our efforts. This goodwill of the greater community is a tangible force and is certainly a key to the profitability and livelihood of the farm. It is one of the great joys of farming. It is also the traditional foundation for a mutually supportive, caring society. This basic concept is often overlooked yet is of critical importance to the success of the small-scale farming operation. It is dependent upon a clear objective of providing for others in a manner that works in harmony with nature.

To engage in the pursuit of providing for others is also of great importance to the drive and will to carry through on difficult agricultural conditions. If one is engaged in agricultural pursuits for self-interest only, it is less likely that there will be a sustained motivation to provide for customers. However, if the motive is a desire to provide for others, great efforts are often exerted. This has definite ramifications for profitability. If one is harvesting and washing turnips in the washing shed on a 35°F (2°C) day solely for one's own profit, it is probably an unhappy situation and likely easily abandoned. If, however, the objective is to provide for others, the uncomfortably chilly conditions will be tolerated and perhaps even laughed at, with the end result being more vegetables for more customers. Customers notice and appreciate consistency and effort, and their loyalty intensifies as a result. This is just a simple example but it reflects daily occurrences that can yield profound results. Loyalty cannot be expected, however. Loyalty is gained and granted only out of free will. If loyalty is expected free will would be diminished, which would then limit loyalty. For loyalty to be at its highest, a philosophy of freedom is critical.

The combination of the drive to provide for others and the understanding of how to work with the natural forces involved in vegetable production is a powerful condition for the development of the vegetable grower. The wisdom gained can go a long way toward stabilizing society and increasing human sensibility. This is often apparent whenever large numbers of people have a close connection to the land. So again, through the efforts to provide for others great gain is provided to the individual grower. Paying attention to the lessons provided accumulates wisdom in the grower. This allows for more effective decision making, and greatly assists in meeting the fears of the individual. These skills are paramount for effective outcomes in agricultural endeavors! It is difficult to stress how important it is to keep our eyes open to the conditions around us. Some of the common barriers to being able to clearly observe are fear, distraction, and stupefaction. These barriers can dramatically impact production and profitability by limiting the information needed for decision making. To help address these barriers, an understanding of the spiritual nature of our existence can be very useful. Indeed, it is difficult to emphasize how important it is to eliminate or otherwise greatly reduce these barriers in order to effectively function on our spiritual journey as growers.

Vegetable growing has traditionally been a very labor-intensive endeavor. As growers it is important for us to be effective at working with those who are assisting in the efforts to get these crops produced. These other people may be family members, interns, apprentices, other growers, volunteers, or farmhands, but generally the same skills and efforts are useful to consider in all cases. To begin, consider the overall goal of the farm. If the goal is primarily health and happiness and the sharing of the bounty of harvest, through the establishment of a business or other entity, then this becomes a powerful force for inspiring others to work to the best of their abilities. It is unlikely that growers can offer a high enough level of pay to be as motivating as a sense of purpose. Indeed, sharing an appropriate sense of purpose may be the best way a grower can motivate others to be as productive as they possibly can. When people all around are doing the best they possibly can, there can be great camaraderie, happiness, and the feeling of shared effort. This often aligns with the grower's primary goals.

Some people are limited in their physical abilities, but in general the shared effort of raising healthful crops helps to improve

FIGURE 15.1. Amid the many tasks of the workday, there is also benefit in allowing time to see and appreciate the beauty of the work the grower has done with natural systems.

physical condition, as well as drive and willpower. Many people who were not raised in an agricultural setting also require much basic instruction to contribute well to the work. Offering steady information for such people is a primary topic of conversation in the field. As they gain more knowledge, the individuals can begin to assemble a more complete picture for themselves of what is going on or, better yet, what needs to go on. Thus farm banter has great purpose.

The growing environment is of course a perfect place for the spiritual development of those who work with us as we interact with the natural forces that are so well prepared to enlighten us. This greatly assists with helping individuals come to terms with their fears, as exposure to the greater forces of nature can bring a more holistic perspective. People's fears are often the primary barrier to them functioning to the best of their abilities. Not that appropriate caution is not in order, but it is very important for individuals to be able to see what is holding them back, and then find the courage to overcome their fears. These conditions of human development can lead to very dedicated people to work with, often to great reward. However, all people are of course free, and only the individual knows their appropriate course at any given time, so flexibility when working with others is again very important.

As growers, we are in the position of being able to assist others in achieving the goal of being able to work as fast as they can. The grower is essentially a coach, similar to a martial arts instructor or a music instructor. The coach shares their experience in order to help others improve performance. The student wants to improve; the coach tries to assist in this effort. This is a vastly improved dynamic over the boss telling a worker what to do. It shows a level of mutual respect for both individuals and creates humorous and harmonious work environments.

Almost all individuals can work to improve efficiency, and for growers it is important to identify talents and help people work on their deficiencies. This often requires tolerance and compassion. A concept that seems important for growers to understand is that not only has the environment become extremely damaged, particularly over the last few decades, but humans have as well. There is widespread physical and mental damage in the human population resulting from the tremendous pollution and imbalance in the modern world. Agricultural effort is likely to engage more of this population as conditions deteriorate.

A greater understanding of our roles among our fellow humans can be gained through understanding of the spiritual nature of humans and the environment. Astrology can be a particularly useful tool for understanding the different perspectives of humans, and can greatly assist growers in identifying strengths and potential weaknesses of those they are working with. The study of the stars can bring much in the way of understanding of not only the interactions of natural forces but the interactions of humans as well.

From a spiritual perspective all existence is interconnected and one action affects another in an ever-widening impact. From this perspective we see the importance of being clear, effective, and intentional with *every and all* actions and decisions in an agricultural endeavor. If the actions of the grower are consistent and allied with natural forces for the proliferation of life, then the ramifications of the waves of interconnectedness bring ever-greater support and alignment for agricultural success. If the actions of the grower are inconsistent, then the waves of ramifications can bring opposition, disorganization, and challenge. The world of nature and spirit is constantly attuning to our actions and intent, and it is never fooled. The "green-thumbed" individual is not necessarily one who is agriculturally trained. Rather it is the one who has the state of mind and spirit supported by life-benefiting intentions.

The natural world's depth of interconnections is often referred to as the web of life, and there are many examples presented to growers. The roots of plants intertwine and exchange nutrients and information in collaboration with the soil microbiology. Fungal networks stretch throughout the soil and provide for the exchange of information, like a nervous system for the living soil. Plants give off volatile compounds into the air and soil to signal other organisms regarding significant conditions. Another example is how the natural world is full of antennae, whether they are on an insect, or are the hairs on a plant leaf. These sensory gathering organs are monitoring the environment, including the grower's actions and intentions. The natural world is thoroughly interconnected, sentient, and actively communicating with itself. As humans we are somewhat cut off from this interconnection. This has given us the ability to develop as we have, but when it comes to agricultural efforts, it is very important to deeply consider the interconnected condition of crop plants and their surroundings. This has definite ramifications for companion planting, crop rotation, management of nutrients, disease, and insects, and much more.

Observation of the interconnected natural world is a primary skill for growers to develop. This is done by spending as much time as possible in contact with nature and paying close attention to surroundings. Regular field walks are very useful, almost always yielding productive observations. It is often said, "The best fertilizer is a farmer's footsteps." However, observation of the field is not the only place to deepen understanding of natural systems; much that is learned in other natural settings can be also of great use. There are many useless distractions in our modern time. It may be wise to be wary of conditions and situations that significantly limit one's ability to directly observe nature.

The growing of vegetables requires constant grower readjustments in the moment to make the best out of any given condition. These conditions, especially in the modern environment, change constantly and sometimes dramatically. Conditions for the growth of vegetables are increasingly unstable. At the same time the world of physical matter and particularly the state of mind of humans has become increasingly rigid. Rigidity is brittle, is weak, and snaps in the winds of change. At this point growers generally benefit from bringing flexibility into themselves as well as their crops. It depends on the skill of the individual grower how much planning is required, and how much flexibility can be integrated. Planning is critical to a grower's efforts, yet too much planning can be a waste of time, producing too much rigidity and lacking the flexibility to allow for making the best call in the moment. With too much flexibility, chaos easily ensues. For instance, you get up on a sunny morning, make a plan for the day, but then the weather changes to rain. Rigidity would have us slog through with the plan—but is there another task that would be better suited to the change in the weather and a more efficient use of the time given wet conditions? Efficiency is highly improved by both planning and flexibility. Maybe the most efficient action is to slog through and finish a project, but quite possibly there is more efficient action to be taken elsewhere. Therefore, it is most efficient to have a plan as well as the flexibility to change plans.

Here are some examples of this kind of approach. When securing accounts (such as a grocery store) and taking on a commitment to provide a certain amount of produce, we plan to produce more than this amount in order to be sure of sufficient supply (overproduce). We then have flexibility in place to also sell excesses (at a farmers market, for instance) if overproduction occurs. We will plan to use a tractor-mounted machine for a certain task, yet if the machine fails, there is the flexibility to utilize hand tools to complete the task. If 100 hours of labor help a week is needed, maybe plan to bring in 120 hours so that there will more likely always be the 100 that is required. We set up a thorough crop production plan and crop rotation yet have flexibility built into it so that variances in weather, labor, and crop health do not disturb the efficient harvest and steady delivery of produce. These are just simple examples of the usefulness of a well-planned yet highly flexible system not only in terms of efficiency and profitability but for the grower's health and happiness as well.

In order to heighten one's ability to observe, the use of trials can be implemented. In trialing, various conditions are set up to be compared with one another. This is often as simple as not doing to a small area whatever one is engaged in doing. For example, when fertilizing, leave one small area unfertilized. When plowing, don't plow a small area. When harvesting, leave a few plants unharvested. In this trial any given action or material can then be compared with a "control" or unaltered area. This kind of trial can pretty much happen constantly in an agricultural setting and is very useful in determining if an appropriate decision or action was made. If the action or material was of benefit, then this can be noted and may lead to further progress. If the control proved to be the better situation, much can be learned from a careful examination of the conditions. Observation skills are critical to spotting subtle changes. Slow concise observation combined with an ability to view the trial in the context of the whole of the interconnections over an extended period of time often gives beneficial insights into appropriate actions.

Aside from simple "leave one spot alone" trials, you can set up other trials to formally compare the effects of multiple materials or actions. Such trials can be rewarding and enlightening, useful for not only assessing new approaches but also reassessing the old ones, because many times things change. Examples of multiple subject trials include comparing different fertilizer materials against each other, or tillage methods, or mulch materials, or crop varieties. When comparing multiple subjects it is often best to keep the layout as consistent as possible in terms of land area, timing, and other variables. Though this approach is similar to scientific trialing, most farm trials are subject to effects by un-isolated variables in a more or less constant state of change. This technically unscientific information is, nonetheless, frequently important to an overall understanding for future success.

Informal trialing can be much more flexible than scientific modeled trials and still provide useful information. An example of general trialing would be to grow a few specific varieties of a crop for several years on different land areas on the farm and then assess which variety proved most successful. If a grower has the time and ability to isolate variables in the trialing, a more scientific approach may also provide useful insights. Growers often benefit from both approaches. However, it is important to remember that truth is a moving target, and what is true at one moment may not be true the next.

An important aspect of trialing is retaining the information gleaned. Although much of the information will simply be kept in the mind of the grower and serve to further develop thinking and abilities, much can be gained from writing down the results of more complex trials. Yield data, crop timing dates for succession plantings, and fertilizer mixture results are common examples of trials that would benefit from written records.

Observation skills and extensive trialing are key aspects to developing appropriate decision-making and are based on the ability of the individual grower. Various off-farm sources also offer advice, such as other growers, consultants, and government advisers. Many ideas can be gathered from such sources that are worthy of trialing on the farm.

Making decisions is a daily, even hourly, part of being a grower. Despite the pressure of the moment, growers do well to strive to make decisions deliberately and not act simply on advice, out of routine, or otherwise blindly. In this regard a grower's intuition can be a great asset. The intuitive sense of an individual is generally a deeply personal voice, the culmination of the various influences that have affected the grower. Much in our technologically dominated world attempts to overcome a person's intuition, because an individual acting on intuition is unlikely to be able to be taken advantage of by the present system of rulership. Thus, it is important for growers to actively develop their intuitive senses. A developed intuition speaks from the heart and the spirit, and it is unlikely to lead one astray.

Along with developing intuitive senses comes the ability to have a closer relationship with the world of the spirit. Much can be gained from paying attention to our dreams, or signs from the spiritual work that surrounds us. In creation all material manifestations are the solidification of spiritual forces. As growers we literally work with the matter of spiritual manifestation. The more open we are to spiritual insights, the more effectively we will be able to work with these divine materials. My hope is that growers will seek the road of compassion, love, kindness, generosity, respect for others, and a sense for the interconnectedness of creation. If these kinds of concepts are embedded in a grower, the likely result is a flexible grower. Flexibility at this time in the earth's development is an essential response to the rampant rigidity of human civilization solidifying around us. Growers need to be able to respond to this altering environment with broad flexibility in order to maintain efficiency. Efficiency starts with making the right call at the right time, which, in turn, requires flexibility. Efficiency continues with the will and determination developed in an attentive grower.

In this chapter an attempt was made to show that agricultural success is based in the mind of the grower; the focus of the mind on naturally harmonious goals provides drive and understanding. Agricultural success then becomes further training for the grower to develop even more fully in an ever-improving loop. When this is occurring the signs are obvious. Most of what is presented here is not new to vegetable growing. It was probably in the realm of "common sense" only a few decades past. But much has been forgotten. It is worth reviewing, even for those well along the road already. It takes the more traditional approach of respect for self, others, and life in general as the key to successful farm management. Not a likely approach to be taught in business schools. However, as we work with creation, the gods' universal plan, the angels, faeries, nature spirits, and one another, we have the gift of being able to build a better future for those who will come after us.

Acknowledgments

This book would not have been possible without the assistance and support of so many people. To all, I express my gratitude. To Anita, my wife and farming partner for all these years, who has given the support and love critical to the creation of this manual. Our daughter Clara, who provided proofreading, cookies, and tolerance for her very busy father. My father, mother, and brother for their hardworking encouragement. Gilbert Risley for mentoring Anita and me and turning a couple of young neighbors into farmers. All of the other growers and farmers who have shared what they understood freely, so that all could improve. The many people who have assisted in the farm's labor so that these methods could develop out of our group effort. The typists who deciphered my handwritten manuscript, and all the bright people at Chelsea Green Publishing. Thank you all.

Further Reading

Brady, Nyle C., and Raymond R. Weil. *The Nature and Properties of Soil.* 15th ed. London: Pearson, 2016.

Brunetti, Jerry. *The Farm as Ecosystem.* 2nd ed. Austin, TX: Acres USA, 2014.

Cho Han Kyu. *Natural Farming.* Seoul, Korea: Cho Global Natural Farming, 2010.

Cho Ju-Young. *Natural Farming: Agriculture Materials.* Seoul, Korea: Cho Global Natural Farming, 2010.

Coleman, Eliot. *The New Organic Grower.* White River Junction, VT: Chelsea Green, 2018.

Datnoff, Lawrence E., Wade H. Elmer, and Don M. Huber, eds. *Mineral Nutrition and Plant Disease.* St. Paul, MN: American Phytopathological Society, 2007.

Dawling, Pam. *Sustainable Market Farming.* Gabriola Island, BC, Canada: New Society Publishers, 2013.

Kains, Maurice G. *Five Acres and Independence.* New York: Dover, 1973.

King, F. H. *Farmers of Forty Centuries.* Mineola, NY: Dover, 1911.

Lovel, Hugh. *Quantum Agriculture.* Blairsville, GA: Quantum Agriculture Publishers, 2014.

Phillips, Michael. *Mycorrhizal Planet.* White River Junction, VT: Chelsea Green, 2017.

Rateaver, Bargyla, and Gylver Rateaver. *The Organic Method Primer.* San Diego, CA: The Rateavers, 1993.

Rodale, J. I. *The Complete Book of Composting.* Emmaus, PA: Rodale, 1960.

Sattler, Friedrich, and Eckard von Wistinghausen. *Bio-dynamic Farming Practice.* Bio-dynamic Agricultural Association, 1989.

Schraefer, Donald. *From the Soil Up.* Austin, TX: Acres USA, 1984.

Steiner, Rudolph. *Agriculture.* Kimberton, PA: Bio-dynamic Farming and Gardening Association, 1993.

Syltie, Paul. *How Soils Work.* Fairfax, VA: Xulon Press, 2002.

Tisdale, Samuel L., and Werner L. Nelson. *Soil Fertility and Fertilizers.* New York: Macmillan, 1966.

Walter, Charles. *Eco-Farm.* Austin, TX: Acres USA, 2003.

Weathers, John. *French Market-Gardening.* London: John Murray, 1909.

Index

Note: Numbers followed by "t" refer to tables, and numbers followed by "f" refer to figures.

About the Author

Bryan O'Hara has been growing vegetables for a livelihood since 1990 at Tobacco Road Farm in Lebanon, Connecticut. Bryan works with natural systems to build complex and balanced soil life, the result of which is a highly productive, vibrant growing system. He uses techniques that include no-till soil management, cover cropping, composting, foliar fertilization, and indigenous microorganism cultures to grow crops year-round. Bryan was named Northeast Organic Farming Association's Farmer of the Year in 2016. He speaks throughout the Northeast and beyond on vegetable production techniques and is known for providing mountains of details in a concise, practical, and cohesive manner.